Service Supply Chain Systems

Communications in Cybernetics, Systems Science and Engineering

ISSN: 2164-9693

Book Series Editor:

Jeffrey Yi-Lin Forrest

International Institute for General Systems Studies, Grove City, USA
Slippery Rock University, Slippery Rock, USA

Volume 8

Service Supply Chain Systems

A Systems Engineering Approach

Editor

Tsan-Ming Choi

Business Division, Institute of Textiles and Clothing,
The Hong Kong Polytechnic University, Hung Hom, Kowloon, Hong Kong

CRC Press
Taylor & Francis Group
Boca Raton London New York

CRC Press is an imprint of the
Taylor & Francis Group, an **informa** business

A BALKEMA BOOK

CRC Press
Taylor & Francis Group
6000 Broken Sound Parkway NW, Suite 300
Boca Raton, FL 33487-2742

First issued in paperback 2018

CRC Press/Balkema is an imprint of the Taylor & Francis Group, an Informa business

© 2016 Taylor & Francis Group, London, UK

No claim to original U.S. Government works

ISBN: 978-1-138-02829-6 (hbk)
ISBN: 978-1-138-61220-4 (pbk)

Typeset by MPS Limited, Chennai, India

Library of Congress Cataloging-in-Publication Data

Names: Choi, Tsan-Ming, author.
Title: Service supply chain systems : a systems engineering approach /
    Tsan-Ming Choi, Business Division, Institute of Textiles and Clothing, The
    Hong Kong Polytechnic University, Hung Hom, Kowloon, Hong Kong.
Description: 1 Edition. | Boca Raton : CRC Press/Balkema, 2016. | Series:
    Communications in cybernetics, systems science and engineering, ISSN:
    2164-9693;Volume 8 | Includes bibliographical references and index.
Identifiers: LCCN 2016000609 (print) | LCCN 2016012255 (ebook) | ISBN
    9781138028296 (hardcover : alk. paper) | ISBN 9781315682143 (ebook)
Subjects: LCSH: Customer services. | Customer relations. | Business logistics.
Classification: LCC HF5415.5.C4784 2016 (print) | LCC HF5415.5 (ebook) | DDC
    658.7–dc23
LC record available at http://lccn.loc.gov/2016000609

Published by: CRC Press/Balkema
                P.O. Box 11320, 2301 EH Leiden, The Netherlands
                e-mail: Pub.NL@taylorandfrancis.com
                www.crcpress.com – www.taylorandfrancis.com

**Visit the Taylor & Francis Web site at
http://www.taylorandfrancis.com**

**and the CRC Press Web site at
http://www.crcpress.com**

Table of contents

Editorial board

Preface

Service supply chain management is a hot topic nowadays. In fact, a service supply chain is a supply network that transfers resources into services or servitised products, with or without physical products, to satisfy customer needs. Managing service supply chains requires innovative strategies, and the systems engineering approach, which emphasizes applying a multi-disciplinary approach to identify optimal solutions, provides the needed framework to help achieve integrated and coordinated service supply chain systems.

Motivated by the importance of service supply chain management and the applicability of the systems engineering approach, I have compiled this CRC handbook which features a few important and interesting research papers as shown below:

- Service supply chain management – A systems engineering approach
- Dynamic coordination of service supply chains
- Signs in service supply
- Supply challenges to management accounting in the new paradigm of service
- Managing service supply chains in the big data era: A system of systems perspective
- Evaluation of discrete event simulation software to design and assess service delivery processes
- The impact of power structure on service supply chain management
- Resilience and complexity in a maritime service supply chain's everyday operation
- Fast fashion retail operations services: An empirical study from consumer perspectives

All of the above papers are peer-reviewed.

As a systems engineer myself, I am very pleased to find that this book has examined service supply chain systems from a lot of different perspectives. This truly reveals the spirit of systems engineering in which a problem should be explored from different angles and the final collaborative effort will lead to the best possible solution.

Before ending, I would like to take this opportunity to express my hearty thanks to *Jeffrey Yi Lin Forest* and *Alistair Bright* for their kindest support for this important book project. I am grateful to all the authors who have contributed their interesting research to this book. I am also indebted to the reviewers for their timely and constructive review comments.

Tsan-Ming Choi
The Hong Kong Polytechnic University
December 2015

About the editor

Tsan-Ming Choi (Jason) is currently a Professor of Fashion Business at The Hong Kong Polytechnic University. Over the past few years, he has actively participated in a variety of research projects on supply chain management and applied optimization. He has authored/edited over twelve research handbooks and guest-edited twenty special issues for various leading journals on related topics. He has published extensively in peer-refereed academic journals such as Annals of Operations Research, Automatica, Computers and Operations Research, Decision Support Systems, European Journal of Operational Research, IEEE Transactions on Automatic Control, IEEE Transactions on Automation Science and Engineering, IEEE Transactions on Cybernetics, IEEE Transactions on Engineering Management, IEEE Transactions on Industrial Informatics, IEEE Transactions on Systems, Man, and Cybernetics, International Journal of Production Economics, International Journal of Production Research, Journal of the Operational Research Society, Omega, Naval Research Logistics, Production and Operations Management, Service Science (INFORMS Journal), Supply Chain Management – An International Journal, Textile Research Journal, Transportation Research – Part E, etc. He is currently an area editor/associate editor/guest editor of Annals of Operations Research, Decision Support Systems, IEEE Systems Journal, IEEE Transactions on Industrial Informatics, IEEE Transactions on Systems, Man, and Cybernetics – Systems, Information Sciences, Production and Operations Management, Risk Analysis, Transportation Research – Part E, and various other operations management and systems engineering journals. He served as an executive committee member/officer of professional organizations such as IEEE-SMC(HK) and POMS(HK) for a few years. He received the President's Award for Excellent Achievement of The Hong Kong Polytechnic University in Nov, 2008 (the most prestigious award for a faculty member at the university level). He was named as the distinguished alumnus of the Department of Systems Engineering and Engineering Management, Faculty of Engineering, The Chinese University of Hong Kong, during the Faculty's 20th Anniversary in 2011. He received the Best Associate Editor Award of IEEE SMC Society in two consecutive years 2013 and 2014. Most recently, he and his collaborators received The Second

Prize of Natural Science Award of Scientific Research Excellence (Science and Technology) in Colleges and Universities, by Ministry of Education of China, on a research theme entitled the "Theory, Method, and Application of Game under Uncertain Environments" in December, 2015. Before joining his current department in fall 2004, he was an assistant professor at The Chinese University of Hong Kong. He is a member of various internationally renowned professional organizations such as IEEE and INFORMS.

Contributing authors

Chen, Xu
School of Management and Economics, University of Electronic Science and
Technology of China, Chengdu, 611731, P. R. China
xchenxchen@263.net

Choi, Tsan-Ming
Business Division, Institute of Textiles and Clothing, The Hong Kong Polytechnic
University, Hung Hom, Kowloon, Hong Kong
jason.choi@polyu.edu.hk

Chow, Pui-Sze
School of Management, Centennial College, 3 Wah Lam Path, Pokfulam, Hong Kong
lpschow@outlook.com

Cinquini, Lino
Institute of Management, Scuola Superiore Sant'Anna
Piazza Martiri della Libertà, 24, 56127 Pisa, Italy
l.cinquini@sssup.it

Ivanov, Dmitry
Department of Business Administration, Berlin School of Economics and Law,
Berlin, Germany
divanov@hwr-berlin.de

Ivanova, Marina
Department of Business Administration and Economics, Chemnitz University of
Technology, Chemnitz, Germany
drivanovmarina@gmail.com

Kataria, Aditi
Maritime Risk and System Safety (MaRiSa) Research Group
World Maritime University
P.O. Box 500, SE 201 24 Malmö, Sweden
katariaditi@gmail.com

Li, Wing-Yan
Business Division, Institute of Textiles and Clothing, The Hong Kong Polytechnic
University, Hung Hom, Kowloon, Hong Kong
winterjoey2002@hotmail.com

Löbler, Helge
Institute for Service und Relationship Management ISRM, Universität Leipzig, Leipzig, Grimmaische Str. 12, 04109 Leipzig, Germany
loebler@wifa.uni-leipzig.de

Pezzotta, Giuditta
CELS – Research group on Industrial Engineering, Logistics and Service Operations
DIGIP – Department of Management, Information and Production Engineering
Università degli Studi di Bergamo, viale Marconi 5, 24044, Dalmine (BG), Italy
giuditta.pezzotta@unibg.it

Pinto, Roberto
CELS – Research group on Industrial Engineering, Logistics and Service Operations
DIGIP – Department of Management, Information and Production Engineering
Università degli Studi di Bergamo, viale Marconi 5, 24044, Dalmine (BG), Italy
roberto.pinto@unibg.it

Pirola, Fabiana
CELS – Research group on Industrial Engineering, Logistics and Service Operations
DIGIP – Department of Management, Information and Production Engineering
Università degli Studi di Bergamo, viale Marconi 5, 24044, Dalmine (BG), Italy
fabiana.pirola@unibg.it

Praetorius, Gesa
Maritime Risk and System Safety (MaRiSa) research group
World Maritime University
P.O. Box 500, SE 201 24 Malmö, Sweden
gesa.praetorius@gmail.com

Rondini, Alice
CELS – Research group on Industrial Engineering, Logistics and Service Operations
DIGIP – Department of Management, Information and Production Engineering
Università degli Studi di Bergamo, viale Marconi 5, 24044, Dalmine (BG), Italy
alice.rondini@unibg.it

Sokolov, Boris
Intelligent Systems Lab, Saint Petersburg Institute for Informatics and Automation of the RAS (SPIIRAS), St. Petersburg, Russia
sokol@iias.spb.su

Tenucci, Andrea
Institute of Management, Scuola Superiore Sant'Anna,
Piazza Martiri della Libertà, 24, 56127 Pisa, Italy
a.tenucci@sssup.it

Wang, Xiaojun
School of Economics, Finance and Management, University of Bristol, Bristol, BS8 1TZ, UK
xiaojun.wang@bristol.ac.uk

Part I

Introduction

Chapter 1

Service supply chain management – A systems engineering approach

Tsan-Ming Choi

Business Division, Institute of Textiles and Clothing, The Hong Kong Polytechnic University, Hung Hom, Kowloon, Hong Kong

SUMMARY

Service supply chain management is an emerging and critical area. As an introductory chapter, we first examine how the systems engineering approach, which includes four critical steps (defining systems requirements, designing and developing systems, operating the systems, and supporting the systems), can be applied for exploring service supply chain systems. Then, we introduce the papers featured in this book and propose how they could fit into the systems engineering approach.

Keywords

Service supply chain systems, service supply chain management, systems engineering approach, systems science, systems analysis and design

1.1 INTRODUCTION

A system can be defined as set of interrelated components working together to achieve some common objectives (Blanchard and Fabrycky 1990). To support service operations in industries such as banking, fashion, healthcare, tourism, etc, dynamic systems (or systems of systems) in the form of service supply chains are established (Wang et al. 2015). To properly achieve the best service supply chain requires substantial efforts and scientific studies (Stavrulaki and Davis 2014).

Systems engineering, a term which was first used by the renowned Bell Laboratories in the 1940s (Brill 1998) and publicized by the pioneering texts such as Goode and Machol (1957) and Hall (1962), can be defined as an interdisciplinary approach and means to enable the realization of successful systems (INCOSE 2015). Systems engineering focuses on the whole instead of the parts (Ackoff 1971, Booton and Ramo 1984). It puts strong emphasis on the integration of multiple disciplines and efforts in order to achieve systems optimality. Systems engineering is a management technology (Sage 1995) and it helps to control the definition, development and deployment of a system in a way to satisfy user requirements.

To explore service supply chain systems, one can adopt the *systems engineering*[1] *approach*. Here, the systems engineering approach can be viewed as one which aims to achieve the effective and efficient system by accomplishing a series of steps from the point when a need is identified to the point when the system is used by consumers. Undoubtedly, the systems engineering approach matches very well with the goal of service supply chain management which is the optimization of the whole service supply chain system with the focal point on satisfying customer needs (Choi et al. 2016). In the following, we discuss the steps involved in the standard systems engineering approach (Blanchard and Fabrycky 1990) for service supply chain management.

Step 1. Defining systems requirements: The whole systems engineering approach starts from identifying the need from "consumers". The consumers here refer to the users who can be individual people or companies, and they are the parties which will consume the services provided by the service supply chain system. In fact, the goal of any supply chain systems, including both "traditional product based" supply chains and service supply chains, is to satisfy the customer requirements. In addition to the identification of consumer need, Step 1 also involves the determination of technological requirements for the service supply chain system (Sage 1998). As a remark, it is important to realize that user requirements in general are variable and the reasons accounting for the respective volatility can be found in Pena and Valerdi (2015).

Step 2. Designing and developing the systems: Based on the systems requirements identified in Step 1, we proceed to create the specific service systems design. To be specific, following the proposal by Blanchard and Fabrycky (1990), we may complete the service systems design step by first creating the conceptual service systems design, then the preliminary service systems design, and finally the detailed service systems design. Some specific related tasks include: (i) Determining the major operations support function, (ii) defining performance metrics, (iii) specifying the systems development process, (iv) selecting suppliers and partners, (v) preparing the needed contractual arrangements for service deployment.

Step 3. Operating the systems: After the systems design and development step is completed, the service systems would be put into applications. It is important to design the service systems coordination schemes and impose the performance testing mechanisms to ensure the proper running of the service supply chain systems. The key point of Step 3 is to ensure the systems requirements are achieved and the systems design elements are all maintained throughout the operations.

Step 4. Supporting the systems: No systems are perfect and they need to be fine-tuned, maintained and enhanced from time to time. We thus have Step 4 which is devoted to providing the needed tasks to help improve the service systems and to provide the needed support to ensure the sustainable operations of the corresponding service supply chains (Filippi et al. 2015). One critical task in this step is to assess customer services and to have continuous service quality assessment of the systems in the consumer environments (Cho et al. 2012).

In this book, we explore different facets of service supply chain systems in two major categories, namely the theoretical advances, and the industrial applications

---

[1]For the development and definitions of modern systems engineering, please refer to Brill (1998).

and cases. Each category features some carefully selected and peer-refereed academic research papers. In the following, we briefly introduce them one by one.

1.2 THEORETICAL ADVANCES

First of all, in Chapter 2, Ivanov et al. theoretically investigate the dynamic coordination of service supply chain systems. They introduce a dynamic model and examine how it can be employed to model service supply chain problems. The authors employ the optimal control theory to tackle the service supply chain coordination problem. They also discuss the practical implementation of feedback control with the use of RFID systems.

In Chapter 3, Lobler examines "signs" in service supply systems. There is no doubt that signs appear everywhere. The author argues that signs would render services in many different ways. Based on this argument, he investigates how signs affect services for the respective cases when they carry meaning, when they change our thinking, and when they lead to actions. The author concludes the study by pointing out the critical implications behind signs which are critically important for future service supply chain management.

In Chapter 4, Cinquini and Tenucci explore the main challenges for management accounting in the domain of services. The authors propose six important dimensions of business service-related changes in which the Goods-Dominant Logic perspective differs from that of the Service-Dominant Logic. Based on their findings, they identify many critical yet under-explored areas, such as drivers of profitability, role of costing, and value measurement in a relationship-based construct, which are all challenging to service-oriented management accounting research. The authors conclude by discussing some most important trends and areas in service-related management accounting which deserve future research.

In Chapter 5, from the system of systems perspective, Choi examines service supply chain management in the big data era. The author first describes the features of service supply chains in the big data era. Then, he checks the literature and identifies five critical characteristics of system of systems and proposes that the service supply chain in the big data era is in fact a system of systems. Based on this argument, he continues to examine the technologies which can support service supply chain operations in the big data era and propose some key principles to enhance service supply chain management. He also discusses some promising future research directions.

1.3 INDUSTRIAL APPLICATIONS AND CASES

Noting that discrete event simulation software is pertinent to the success of service delivery processes in practice, Pezzotta et al. study in Chapter 6 the evaluation scheme of some commercial discrete event simulation software systems. The authors conduct a systematic comparative analysis among the four selected simulation software systems. With the collected real case data, the authors present the detailed analysis on how the discrete event simulation software can be evaluated. They also select the best candidate for their specific case.

In Chapter 7, Wang and Chen study the impact of power structure on service supply chain systems. The authors present two case studies, one on a retail service supply chain system with the offline and online dual channels, and one on a mobile phone service supply chain. They conduct a game theoretical study to analytically explore the impact of different power structures on the operational decisions on pricing and channel selections in the service supply chain system. For the first case on the retail service supply chain, they find that the retailer prefers the integrated pricing policy over the decentralized pricing policy. Furthermore, they reveal that even though individual supply chain agents are better off when they have more power, the whole supply chain is usually not optimal; in fact, the whole service supply chain is optimal when there is a balanced power distribution in the system. For the second case study, the authors derive a mechanism in which the supply chain agents can employ to develop optimal pricing and subsidy policies to maximize their profits.

In Chapter 8, Praetorius and Kataria investigate the maritime service supply chain system. To be specific, the authors focus on studying the vessel traffic service, which is a central service in guaranteeing traffic flow safety and efficiency in maritime operations. They employ the functional resonance analysis method to develop a functional vessel traffic service model. They conduct the system design analysis and reveal whether the respective service supply chain can operate in a resilient manner. This chapter has good implications for the use of a function-based approach for both maritime service supply chain management and other service operations.

In Chapter 9, Li et al. study the retail operations services in fast fashion supply chain systems. The authors explore the problem by conducting a consumer survey. Based on scientific statistical analysis, they uncover how the quick response service, the enhanced design service, and the pricing strategy of fast fashion retailers affect the customer involvement and purchasing behaviors. They show that a higher level of consumer satisfaction with fast fashion services would lead to a higher level of customer involvement. They further reveal that a higher level of customer involvement is beneficial to the fast fashion retailers.

1.4 CONCLUDING REMARKS

In Section 1.1, we have discussed how the systems engineering approach can be applied for service supply chain management analysis. In Sections 1.2 and 1.3, we have further introduced and presented the papers featured in this book. As a concluding remark, we summarize how the featured papers are related to the four steps in the systems engineering approach in Table 1.1.

From Table 1.1, we can see how different methods can be applied, in different steps of the systems engineering approach, to explore service supply chain problems. As a remark, undoubtedly, no single method can be applied to complete all steps and it is also not necessary for a piece of research to report findings from all the required steps in the systems engineering approach for a service supply chain management problem. Nevertheless, it is important to realize the existence of these four critical steps under the systems engineering approach. For completeness, service supply chain systems studies should pay full attention to them.

Table 1.1   Four Steps of the Systems Engineering Approach, Related Studies and Corresponding Tools, Models and Scopes.

	Related Studies	*Corresponding Tools, Models and Scopes*
Step 1. Defining systems requirements	Li et al.	Service requirements of customers
Step 2. Designing and developing the systems	Ivanov et al.	Optimal control for services
	Lobler	Role played by signs on services
	Cinquini and Tenucci	Management accounting for services
	Praetorius and Kataria	The functional resonance analysis method
	Li et al.	Regression analysis on the impact of different service dimensions
Step 3. Operating the systems	Ivanov et al.	The feedback control with the use of RFID systems
	Pezzotta et al.	Evaluate the discrete event simulation software
	Wang and Chen	Game theoretical analysis and optimization
Step 4. Supporting the systems	Pezzotta et al.	Discrete event simulation software
	Choi	Principles to improve performance of the service supply chain system of systems

REFERENCES

Arkoff, R.L. Toward a system of systems concept. *Management Science*, 17, 661–671, 1971.

Blanchard, B.S., W.J. Fabrycky. *Systems Engineering and Analysis*. Prentice Hall, 2nd edition, 1990.

Booton, R.C., S. Ramo. The development of systems engineering. *IEEE Transactions on Aerospace and Electronic Systems*, AES-20(4), 1984.

Brill, J.H. Systems engineering – a retrospective view. *Systems Engineering*, 1(4), 258–266, 1998.

Cho, D.W., Y.H. Lee, S.H. Ann, M.K. Hwang. A framework for measuring the performance of service supply chain management. *Computers & Industrial Engineering*, 62, 801–818, 2012.

Choi, T.M., Y. Wang, S.W. Wallace. Risk management and coordination in service supply chains: information, logistics and outsourcing. *Journal of the Operational Research Society*, forthcoming, 2016.

Goode, H., R. Machol. *Systems Engineering*. McGraw Hill, New York, 1957.

Filippi, M., A. D'Ambrogio, M. Lisi. A service systems engineering framework with application to performance based logistics. Conference *Proceedings of The 2015 IEEE International Symposium on Systems Engineering (ISSE)*, Sep 2015, Rome, 311–317, 2015.

Hall, A.D. *A Methodology for Systems Engineering*. Van Nostand, Princeton, NJ, 1962.

INCOSE, What is Systems Engineering? http://www.incose.org/AboutSE/WhatIsSE, retrieved on 2 December, 2015.

Pena, M., R .Valerdi. Characterizing the impact of requirements volatility on systems engineering effort. *Systems Engineering*, 18(1), 59–70, 2015.

Sage, A.P. *Systems Management for Information Technology and Software Engineering*. Wiley, New York, 1995.

Sage, A.P. Systems engineering: purpose, function, and structure. *Systems Engineering*, 1(1), 1–3, 1998.

Stavrulaki, E., M.M. Davis. A typology for service supply chains and its implications for strategic decisions. *Service Sciences*, 6, 34–46, 2014.

Wang, Y., S.W. Wallace, B. Shen, T.M. Choi. Service supply chain management: A review on operational models. *European Journal of Operational Research*, 247, 685–698, 2015.

Part II

Theoretical advances

Chapter 2

Dynamic coordination of service supply chains

D. Ivanov[1], B. Sokolov[2] & M. Ivanova[3]

[1] Department of Business Administration, Berlin School of Economics and Law,
Chair of International Supply Chain Management, Berlin, Germany
[2] Saint Petersburg Institute for Informatics and Automation of the RAS (SPIIRAS),
St. Petersburg, Russia
[3] Department of Production and Industrial Management, Chemnitz University of Technology,
Chemnitz, Germany

SUMMARY

This chapter introduces the structure dynamics control concept and a dynamic model to coordinate activities in service supply chains. Continuous nature of services in the supply chain makes them similar to continuous production systems. That is why we discuss in this chapter previous applications of control theory to supply chains with continuous flows. Subsequently, structure dynamics control approach is presented. Dynamic coordination of information services, information resources, and business process activities is considered on the scheduling level modeled as an optimal control problem. Practical implementation of feedback control with the help of RFID technology is discussed.

Keywords

service, supply chain, coordination, structure dynamics, control

2.1 INTRODUCTION

This chapter introduces the structure dynamics control concept and a dynamic model to coordinate activities in service supply chains (SC). Services differ from physical goods in many aspects and can be encountered in finance, healthcare, energy and telecommunications. One of the specific service SC characteristics in comparison to classical manufacturing SCs is that the production process is not divided into discrete batches and processing steps but it is rather continuous. As such, it becomes sensible to apply continuous optimization and control theory (CT) to the coordination analysis in the service SCs.

Dynamic coordination of the service SCs consists of the process alignment in different structures, i.e., information services (IS), information resources (IR), and business process activities (see Fig. 2.1). Most of the new service SC concepts share attributes of smart networking. That is why it becomes a timely and crucial topic to consider such SCs as collaborative cyber-physical systems (Camarinha-Matos and Macedo 2010, Zhuge 2011, Ivanov et al. 2014). Cyber-physical systems incorporate elements from

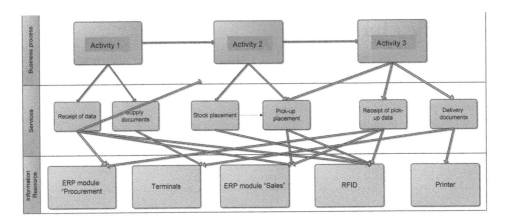

Figure 2.1   Multi-structural view of service coordination (based on Ivanov et al. 2014).

both information and material subsystems which are integrated and decisions in them are cohesive (Zhuge 2011). In addition, such systems evolve through adaptation and reconfiguration of their structures, i.e. through structure dynamics (Ivanov et al. 2010, Ivanov and Sokolov 2012b). These issues are becoming more and more important in practice since the impact of IT on the business processes has become crucial in recent years (Choi et al. 2002, Camarinha-Matos and Maseda 2010, Cannella et al. 2014). Recent research indicated that an aligning of business processes and information systems may potentially provide new quality of decision-making support (Dedrick et al. 2008, Jain et al. 2009).

Such a framework is based on recent developments in cloud computing, see for example studies by Wang et al. (2010) and Jiang et al. (2012). For business process scheduling and control, some ISs are needed. They should be available when material flow is scheduled and executed. The ISs are provided by some distributed IRs which may be subject to full or partial unavailability due to planned upgrades or unpredicted disruptions. Therefore, such an open SN has to be considered as a dynamic system.

Consider a practical example of the integration of business processes in information services in the SC. Lenovo's SC strategy uses a mix of in-house production facilities and outsourcing partners, such as original design manufacturers (ODMs) and electronic manufacturing services (EMS) providers (Pershke 2012). A key decision for any OEM is if they should receive finished goods from the ODM or EMS supplier, and then ship to its customers? Or should they ship directly from the outsourcing partner to the customer? By establishing a unified, global integration platform, they eliminated some "information black holes" that limited the ability to serve the customers. This improvement has enabled the company to implement a new process, called "sell on the water" that delivers product to the hands of the customers faster. With this process, a ship can be loaded at the point of origin with product based on forecasted demand, and then can be seen as a "virtual warehouse". This makes it possible to begin allocating product to customers while the ship is in transit. When a shipment is loaded on board a vessel, this event triggers a receipt into a virtual warehouse location in the U.S. This allows allocation of customer orders against the expected stock up to six weeks sooner than actual physical receipt of the goods in the warehouse.

Under that strategy, Lenovo is using IBM's unified integration platform. They developed a standardized methodology for on-boarding ODMs and EMS providers, both for those using electronic data interchange (EDI) connections and those using the web portal. One ODM that produces Lenovo notebook PCs and uses the EDI Direct Ship enablement solution cut its cost per box by 1%, reduced its order cycle time by five working days and improved quality by avoiding a second touch (Pershke 2012). With the help of automatically generated reports and purchasing information updated daily by ODMs, the SC visibility has been improved. Once Lenovo receives the dispatch advice (ASN) from the ODM, it immediately sends this electronic information to the carrier, along with the routing guide and consolidation instructions. The latter two are fairly static, and are selected when the ASN is sent. From this point, Lenovo has visibility of the shipment all the way through to delivery at the customer's dock. Lenovo has about 100 customers that place orders electronically via EDI.

In recent years, studies on SC dynamics were broadened by developments in information technologies such as RFID (Radio Frequency Identification), SCEM (Supply Chain Event Management) and mobile business which provide a constructive basis to incorporate the stages of SC planning and execution control (Lee and Oezer 2008, Choi 2011).

Considering the case-study and recent research trends, the objective of this chapter is to discuss the applicability of control theory to modeling service SC coordination and exemplify it on an example. The remainder of this chapter is organized as follows. In Section 2.2, applicability of control theory to service SC coordination is discussed. Section 2.3 presents methodology of structure dynamics control. Sections 2.4 and 2.5 develop mathematical model for joint scheduling of business processes and information services. The paper is concluded by summarizing the main findings and discussing future research.

2.2 REVIEWS OF CONTROL THEORY APPLICABILITY TO SERVICE SUPPLY CHAIN COORDINATION

Continuous nature of services in the SC makes them similar to continuous production systems. That is why we discuss in this chapter previous applications of CT to SCs with continuous flows (Ivanov et al. 2012). Dynamics in SCs can refer to both the dynamics of a process under optimization (dynamics of the transition from an input to an output state) and the real-time dynamics regarding the feedback-loop consideration and adaptation in accordance to an actual execution environment. Employment of CT can be favorable for both the SC synthesis and analysis stages.

SCs in continuous production systems comprise multi-stage network of suppliers. Continuous flow scheduling problems have their place in many industries such as gas, oil, chemicals, glass and fluids production as well as production of granular goods and steel details (Shah 2004, Puigjaner and Lainez 2008, Subramanian et al. 2013, Ivanov et al. 2013, Ivanov et al. 2015). Sarimveis et al. (2008) underline the resemblance of SCs to engineering dynamic systems.

Previously, CT has been extensively applied for management and economics applications (Sethi and Thompson, 2006). The applied tools vary from standard transfer function analysis to model predictive control. Optimal design, planning and scheduling is another application of CT, and especially optimal program control (OPC), to the SCM. OPC is a method for solving dynamic optimization problems, when those problems are expressed in continuous time and the values of goal criteria vector J are accumulated over time subject to changes of a state vector x(t) under the influence of constrained control actions u(t). OPC is a deterministic control method as opposed to the stochastic optimal control. One of the basic milestones in modern OPC, along with dynamic programming (Bellmann 1972), is the maximum principle that was developed in the 1950s by Russian mathematicians (Pontraygin et al. 1964).

Pontraygin's maximum principle is an original method for computing OPC when optimizing system behavior over many time periods under constrained control, subject to several decision variables where other techniques can become analytically and computationally difficult to apply. The initial formulation of maximum principle was for the problem of transferring a space vehicle from one orbit to another with minimum time and minimum fuel consumption. Maximum principle basically generalizes the calculus of variations and builds the basis of the OPC theory. The development of the maximum principle has stimulated the application of OPC to many industrial and engineering applications.

Another fundamental aspect of SCM is an analysis of randomness, disturbances and fluctuations. Basically, there are three main properties of an SC which can be analyzed regarding uncertainty. These are: (1) ability to cope with volatility and continue plan execution after being perturbed, (2) ability to remain stable and achieve the planned performance in the presence of disturbances, and (3) ability to maintain, execute and recover (adapt) the planned execution along with the achievement of the planned (or adapted, but yet still acceptable) performance. In the systems and control theories, these properties are analyzed as stability, robustness, and resilience (Ivanov and Sokolov 2013).

Although stability analysis is a useful tool, it is subject to many restrictions if applied in the classical form of Lyapunov's stability or BIBO stability. First, these standard models imply natural movement of objects. Second, they typically consider very small deviations of control and output variables. Third, stability analysis can help in estimating SC volatility in any concrete state. But it is not enough to stabilize the SC – the SC should also bring profits; hence, the inclusion of performance considerations (i.e., the robustness analysis) is required as the next step. Finally, classical stability analysis is concerned with funding equilibrium states for mechanical and automatic systems, but not organizational systems.

A very extensive area of CT applications to SCM is related to the adaptation and real-time control. Actually, it is the reference area for many SC scholars and professionals when first discussing CT. A popular technique of SC adaptive planning is the model predictive control (MPC). In MPC, a system model and both current and historical measurements of the process are used to predict the system behaviour at future instants. A control-relevant objective function is then optimized to calculate a control sequence to satisfy the system constraints.

Applications of MPC to multi-echelon production–inventory problems and SCs have been examined previously. Perea et al. (2000) modeled multi-plant, multi-product polymer process through difference equations and schedule optimisation in MPC framework. Braun et al. (2003) developed a decentralized MPC implementation for a six-node, two-product, and three-echelon demand network problem. In the study by Puigjaner and Lainez (2008), a multi-stage stochastic model has been employed. When applying MPC to SCM, the centralized controller and its functions are essential. In technical systems, the controller is a technical device (e.g., a sensor) that adapts the system behavior based on error identification within milliseconds. The controller in the SC is a manager, or more precisely, a number of managers with possible conflicting interests. Even if a deviation in the SC execution was identified (e.g., track delay identification with the help of RFID or SCEM), the MPC controller will not be able to change anything in this situation. Therefore, the models are needed to identify the deviations and notify the SC managers. This is used to estimate the impact of this disturbance on SC performance, and to produce recommendations on adaptations. That is why additional research is needed to analyze the applicability of MPC to human-driven SC adaptations.

The studies by Sarimveis et al. (2008) and Harjunkoski et al. (2014) showed a wide range of advantages regarding the application of CT models. They include, first of all, a non-stationary process view and accuracy of continuous time (Chen et al. 2012, Jula and Kones 2013). In addition, a wide range of analysis tools from CT regarding stability, controllability, adaptability, etc. may be used if a schedule is described in terms of control. Ivanov and Sokolov (2013) develop a CT framework for analysis of SC robustness, resilience, and stability. Wang et al. (2013) develop a distributed scheduling algorithm called a closed-loop feedback simulation approach that includes adaptive control of the auction-based bidding sequence to prevent the first bid first serve rule and may dynamically allocate production resources to operations. Ivanov et al. (2015) apply OPC to distributed SC scheduling in smart factory Industry 4.0.

2.3 STRUCTURE DYNAMICS CONTROL APPROACH

SCs are characterized by a set of interrelated structures such as organizational, functional, informational, financial, etc. Decisions in all the structures are interrelated. Moreover, the structures are subject to changes; hence, SC structure dynamics (Ivanov et al. 2010) is frequently encountered. The dynamic SC characteristics are distributed upon different structures, e.g. organizational (i.e., agile supply structure), functional (i.e., flexible competencies), product-based (i.e., product flexibility), informational (i.e., fluctuating information availability, financial (i.e., cost and profit sharing). This multi-dimensional dynamic space along with the coordinated and distributed decision-making deals with modern SCs as multi-structural active systems with structure dynamics. Besides, the SC execution is accomplished by permanent changes in the internal network properties and the environment.

The SC dynamic characteristics are distributed upon different structures, i.e.:

• organizational structure dynamics (i.e., agile supply structure),
• functional structure dynamics (i.e., flexible competencies),
• information structure dynamics (i.e., fluctuating information availability), and
• financial structure dynamics (i.e., cost and profit sharing).

This multi-dimensional dynamic space along with the coordinated and distributed decision-making leads us to the understanding of the SCs as multi-structural systems with structure dynamics (Ivanov et al. 2010, Ivanov and Sokolov 2012b). The basic ideas of the SC representation as multi-structural dynamic systems are the dynamic decomposition in each of the structure based on the intervals of structural constancy and multi-structural view in each of these intervals.

The main idea is the dynamic interpretation of planning in accordance with the natural logic of time with the help of OPC. Besides, a priori knowledge of the SC structure, and moreover, structure dynamics, is no more necessary – the structures and corresponding functions are optimized simultaneously as the control becomes a function of both states and structures. The splitting of the planning period into the intervals occurs according to the natural logic of time and events. As the approach is based on control theory, it is a convenient approach to describe intangible services due to abstract nature of state variables which can be interpreted as abstract service volumes.

Based on the structure dynamics control, an explicit picture of possible structural combinations becomes evident. Some examples of the structural interrelations follow. Business processes are designed in accordance with SC goals and are executed by organizational units. These units fulfil management operations and use certain technical facilities and information systems for planning and coordination. Business processes are supported by information systems. Organizational units have a geographical (topological) distribution that also may affect the planning decisions. Collaboration and trust (the so-called "soft facts") in the organizational structure do affect other structures, especially the functional and informational structures. Managerial, business processes (distribution, production, replenishment etc.), technical and technological activities incur SC costs, which also correspond to different SC structures.

So the representation of SCs as complex multi-structural dynamic systems can be favourable both for the identifying structures and corresponding models and for identification of different structural relations from the static and dynamic points of view. These relations are important not only at the planning, but also at the execution stage. For example, the planned or unplanned changes in the information system structure (e.g., an update of the IS or a disruption in data transferring or processing) should be matched with the business process structure so that (1) the update/maintenance jobs do not influence the business process execution and the SC performance as well.

Cyber-physical systems are characterized by decentralization and autonomous behavior of their elements. In addition, such systems evolve through adaptation and reconfiguration of their structures, i.e. through structure dynamics that exists both in classical SCs and agile SCs with flexible structures. In these settings, two questions may be raised: (1) what is the optimal volume of information services needed to ensure operation of SCs and (2) how these services shall be scheduled at the planning stage and re-scheduled (adapted) in dynamics at the execution control stage. Conventionally, the above-described two problems are solved step-by-step. With the help of structure dynamics control, a special dynamic representation of multi-structural networks can be developed where such problems can be solved simultaneously.

In addition, due to the increasing role of information services in different forms, e.g., cloud computing, the service-based approaches to integrated planning and scheduling of both material and information flows in SCs are needed. Such integration is also important to prevent failures of IS-enabled SCs. Although recent research has extensively dealt with SC scheduling and IS scheduling in isolation, the integrated scheduling of both material and information flows still represents a research gap.

2.4 TECHNICAL MODEL

We relate to the problem structure in Fig. 2.1 and the work by Ivanov et al. (2014). The SC is modelled as a networked controlled system described through a dynamic interpretation of the operations' execution. The execution of operations is characterized by (1) results (e.g., processed volume, completion time, etc.), (2) intensity consumption of the machines, and (3) supply and information flows resulting from the schedule execution. The operations control model (M1) is first used to assign and sequence ISs to operations in material flows, and then a flow control model (M2) is employed to assign and schedule jobs at IRs subject to the requirements on the ISs availability. The basic interaction of these two models is that after the solving M1, the found control variables are used in the constraints of M2. Note that in the calculation procedure, the models M1 and M2 will be solved in a coordinated manner, i.e., the scheduling problems in all the structures (i.e., material flows, ISs, and IRs) will be integrated.

The basic conceptual idea of this approach is the fact that the operation execution and machine availability are dynamically distributed in time over the planning horizon. As such, not all operations and machines are involved in the decision making at the same time. Therefore, it becomes quite natural to transit from large-size allocation matrices with a high number of binary variables to a scheduling problem that is dynamically decomposed.

Following an approach to decompose the solution space and to use exact methods over its restricted sub-spaces, we propose to use the OPC theory for the dynamic decomposition of the scheduling problem. The computational procedure will be based on a modified maximum principle in the continuous form blended with MP. One of the basic problems in applying any decomposition method is how then to evaluate the overall system performance. By using the maximum principle, this problem is solved a priori. Since the maximum principle guarantees that the optimal solutions of the instantaneous problems give an optimal solution to the overall problem, this principle is a convenient approach to a problem decomposition into a number of sub-problems. For these sub-problems, optimal solutions can be found, e.g., with the help of discrete optimization. Then these solutions are linked into an OPC.

The main advantage of the proposed approach to the problem class is the consideration of real event logic in the flow shop. The flow shop scheduling problem becomes naturally decomposed according to the logic of time. In the existing optimal scheduling methods, time aspects are represented as quite limited. For example, within discrete optimization, for the scheduling horizon, all possible assignments are represented within one system of algebraic equalities or inequalities. In the proposed dynamic interpretation of the execution of the operations, the large-scale multi-dimensional combinatorial matrix is decomposed. If so, the OPC and discrete optimization models can be integrated subject to their advantages. The calculation procedure is transferred to discrete optimization methods and is therefore independent from the OPC. The solution at each point of time is computed with the help of discrete optimization.

OPC is used for modeling the execution of operations and interlinking the discrete optimization solutions over the planning horizon.

2.4.1 Model M1

The *model of operation execution dynamics* can be expressed as Eqs. (2.1)–(2.3):

$$\frac{dx_i^{(v,l)}}{dt} = \varepsilon_{il}(t) \cdot u_{il}^{(v)}(t) \tag{2.1}$$

$$\frac{dy_{il}^{(v)}}{dt} = \eta_{il}(t)[1 - \vartheta_{il}^{(v,1)} - u_{il}^{(v)} - \vartheta_{il}^{(v,2)}] \tag{2.2}$$

$$\frac{dx_{il}^{(v,1)}}{dt} = \vartheta_{il}^{(v,1)}; \quad \frac{dx_{il}^{(v,2)}}{dt} = \vartheta_{il}^{(v,2)} \tag{2.3}$$

Eq. (2.1) describes operation execution dynamics subject to availability of IS described in the matrix function $\varepsilon_{il}(t)$. $u_{il}^{(v)}(t) = 1$ if service $S_l^{(v)}$ is assigned to the operation $D_i^{(v)}$, $u_{il}^{(v)}(t) = 0$ otherwise. Eq. (2.2) represents idle time in the material flow caused by unavailability of the IS $S_l^{(v)}$. Eq. (2.3) represents the dynamics of operation's execution according to precedence constraints.

The control actions are *constrained* as follows:

$$\sum_{i=1}^{k_j} u_{il}^{(v)}(t) \le g_l^{(v)}; \quad \forall l; \qquad \sum_{l=1}^{d_j} u_{il}^{(v)}(t) \le h_i^{(v)}; \quad \forall i \tag{2.4}$$

$$\sum_{l=1}^{d_j} u_{il}^{(v)} \left[\sum_{\alpha \in \Gamma_{v1}} \left(a_\alpha^{(v,l)} - x_\alpha^{(v,l)} \right) + \prod_{\beta \in \Gamma_{v2}} \left(a_\beta^{(v,l)} - x_\beta^{(v,l)} \right) \right] = 0; \quad \forall v \tag{2.5}$$

$$\vartheta_{il}^{(v,l)} \cdot x_{il}^{(v,l)} = 0; \qquad \vartheta_{il}^{(v,2)} \left(a_{il}^{(v,l)} - x_{il}^{(v,l)} \right) = 0; \quad \forall i; \ \forall l \tag{2.6}$$

$$u_{il}^{(v)}(t) \in \{0, 1\}; \quad \vartheta_{il}^{(v)}(t) \in \{0, 1\} \tag{2.7}$$

Constraints (2.4) are assignment problem constraints. They define possibilities of parallel use of many services for one operation and for parallel processing of many operations at one service. Constraints (2.5) determine the precedence relations. Constraints (2.6) interconnect main and auxiliary controls. Equation (2.7) constraints control to be Boolean variables.

The *start and end conditions* are defined as follows:

$$t = t_0^{(j)} : x_i^{(v)}(t_0^{(j)}) = y_{il}^{(v)}(t_0^{(j)}) = x_{il}^{(v)}(t_0^{(j)}) = 0 \tag{2.8}$$

$$t = t_f^{(j)} : x_i^{(v)}(t_f^{(j)}) = a_i^{(v)}; \quad y_l^{(v)}(t_f^{(j)}); x_i^{(v)}(t_f^{(j)}) \in \mathbf{R}^1 \tag{2.9}$$

Eqs. (2.8) and (2.9) define initial and end values of the variables $x_i^{(v)}(t)$, $y_{il}^{(v)}(t)$, $x_{il}^{(v)}(t)$ at the moments $t_0^{(j)}$ and $t_f^{(j)}$. Conditions (2.9) reflect the desired end state. The right parts of equations are predetermined at the planning stage subject to the planned demand for each job.

The *objectives* are defined as follows:

$$\min J_1^{(v)} = \sum_{i=1}^{k_v} \sum_{l=1}^{d_j} y_{il}^{(v)}(t_f^{(j)}) \tag{2.10}$$

$$\max J_2 = \sum_{i=1}^{k_v} \sum_{l=1}^{d_j} \frac{1}{x_{il}^{(v,2)}(t_f^{(j)})} \int_{t_0^{(j)}}^{t_f^{(j)}} \vartheta_{il}^{(v,2)}(\tau) d\tau \tag{2.11}$$

$$\min J_3 = \sum_{i=1}^{k_v} \sum_{l=1}^{d_j} \int_{t_0^{(j)}}^{t_f^{(j)}} [c_{il}^{(v,1)}(\tau) + c_{il}^{(v,2)}(\tau)] \cdot u_{il}^{(v)}(\tau) d\tau \tag{2.12}$$

Eq. (2.10) minimizes losses from the idle time of services. Eq. (2.11) estimates the service level by the volume of on-time completed jobs in the material flow. Eq. (2.12) minimizes total costs of IS.

2.4.2 Model M2

The *model of operation execution dynamics* in the IRs can be expressed as (2.13):

$$\frac{dx_\chi^{(v,l)}}{dt} = \sum_{r=1}^{\rho_v} u_{\chi r}^{(v,l)}; \quad \frac{dx_r^{(v,l)}}{dt} = \sum_{\chi=1}^{S_l} w_{\chi r}^{(v,l)}; \quad \frac{dx_{rS_l}^{(v,l)}}{dt} = \omega_{rS_l}^{(v,l)} \tag{2.13}$$

Eq. (2.13) describes operation's execution dynamic in the IR subject to operation of the IRs and recovery operations in the case of disruptions in the information structure. The control actions are *constrained* as follows:

$$0 \leq u_{\chi r}^{(v,l)} \leq [e_{\chi r}^{(j)}(1 - \vartheta_r^{(p,2)}(t)) + e_{\chi r}^{-(j)} \vartheta_r^{(p,2)}(t)] w_{\chi r}^{(v,l)}; \tag{2.14}$$

$$\sum_{v=1}^{n_j} \sum_{\chi=1}^{S_v} V_\chi^{(v)} \cdot w_{\chi r}^{(v,l)} \leq [V_r^{(j)}(1 - \vartheta_r^{(p,2)}(t)) + \bar{V}_r^{(j)} \vartheta_r^{(p,2)}(t)] \xi_r^{(j,1)}; \tag{2.15}$$

$$\sum_{v=1}^{n_j} \sum_{\chi=1}^{S_v} u_{\chi r}^{(v,l)}(t) \leq [\Phi_r^{(j)}(1 - \vartheta_r^{(p,2)}(t)) + \bar{\Phi}_r^{(j)} \vartheta_r^{(p,2)}(t)] \xi_r^{(j,2)}; \tag{2.16}$$

$$\sum_{r=1}^{\rho_v} w_{\chi r}^{(v,l)} \left[\sum_{\pi \in \Gamma_{\chi 3}} (a_\pi^{(v,l)} - x_\pi^{(v,l)}) + \prod_{\mu \in \Gamma_{\mu 4}} (a_\mu^{(v,l)} - x_\mu^{(v,l)}) \right] = 0; \tag{2.17}$$

$$\sum_{r=1}^{\rho_v} w_{\chi r}^{(v,l)}(t) \leq \psi_\chi; \quad \forall \chi; \qquad \sum_{\chi=1}^{s_l} w_{\chi r}^{(v,l)}(t) \leq \varphi_r; \quad \forall r; \tag{2.18}$$

$$\omega_{rS_l}^{(v,l)}(a_{S_l}^{(v,l)} - x_{S_l}^{(v,l)}) = 0; \tag{2.19}$$

$$w_{\chi r}^{(v,l)} \in \{0, u_{il}^{(v)}\}; \quad \vartheta_r^{(p,2)}(t), \omega_{rsl}^{(v,l)} \in \{0,1\}; \quad \xi_r^{(j,1)}(t); \quad \xi_r^{(j,2)}(t) \in [0,1]. \tag{2.20}$$

With the help of functions $0 \leq \xi_r^{(j,1)}(t) \leq 1$ and $0 \leq \xi_r^{(j,2)}(t) \leq 1$, perturbation impacts on the IR $B_r^{(v,j)}$ can be modelled. Eqs (2.14)–(2.16) constraint information processing at $B_r^{(v,j)}$ before and after the reconfiguration. Constraints (2.17) set precedence relations on information processing operation similar to Eq. (2.5). Constraints (2.18) are related to assignment problem and are similar to (2.4). Eq. (2.19) determines the conditions of processing completion.

The *end conditions* are defined as follows:

$$t = t_0^{(j)} : x_\chi^{(v,l)}(t_0^{(j)}) = x_r^{(v,l)}(t_0^{(j)}) = x_{rs_l}^{(v,l)}(t_0^{(j)}) = 0; \tag{2.21}$$

$$t = t_f^{(j)} : x_\chi^{(v,l)}(t_f^{(j)}) = a_\chi^{(v,l)}; \quad x_r^{(v,l)}(t_f^{(j)}); \quad x_{rs_l}^{(v,l)}(t_f^{(j)}) \in \mathbf{R}^1. \tag{2.22}$$

The *goals* are defined as follows:

$$J_4 = \sum_{r=1}^{\rho_v-1} \sum_{r_1=r+1}^{\rho_v} \int_{t_0^{(j)}}^{t_f^{(j)}} (x_r^{(v,l)}(\tau) - x_{r_1}^{(v,l)}(\tau))d\tau; \tag{2.23}$$

$$J_5 = \sum_{r=1}^{\rho_v} \sum_{\chi=1}^{s_l} \int_{t_0^{(j)}}^{t_f^{(j)}} \delta_{\chi r}^{(v,l)}(\tau) \cdot w_{\chi r}^{(v,l)}(\tau)d\tau; \tag{2.24}$$

$$J_6 = \frac{1}{2}\sum_{\chi=1}^{s_l} \left(a_\chi^{(v,l)} - a_\chi^{(v,l)}\left(t_f^{(j)}\right)\right)^2. \tag{2.25}$$

$$J_7 = \sum_{\chi=1}^{s_l} \sum_{r=1}^{\rho_v} \int_{t_0^{(j)}}^{t_f^{(j)}} \left[c_{\chi r}^{(l,1)}(\tau) + c_{\chi r}^{(l,2)}(\tau)\right] w_{\chi r}^{(v,l)}(\tau)d\tau; \tag{2.26}$$

Eq. (2.23) estimates uniformity of the use of the IRs $B_r^{(v,j)}$ and $B_{r_1}^{(v,j)}$; $r, r_1 \in \{1, \dots, \rho_v\}$. Eq. (2.24) estimates amount of completed operations $D_{<l,\chi>}^{(v,j)}$. Eq. (2.25) takes into account losses from non-fulfilled operations. Eq. (2.26) assesses total cost of ownership (TCO) for the IR $B_r^{(v,j)}$.

The developed modeling complex is composed of dynamic models of IS and IR control subject to execution of material flows. It also includes elements of IR reconfiguration (e.g., in Eqs. (2.14)–(2.16) and (2.20)). The presented models M1 and M2 are interconnected with the help of Eq. (2.6) where elements from M2 are used in M1. In turn, M1 influences M2 through the Eqs. (2.14) and (2.20).

2.4.3 Feedback control implementation with the help of RFID

For the investigation of the RFID-based feedbacks within the previously developed software prototype for dynamic SC scheduling (Ivanov and Sokolov, 2012; Ivanov et al. 2013), an experimental stand with a transportation network and some production and warehouse facilities is currently under development. After the schedules are set up, the stage of SC execution control follows. At the physical level, cargo movement control takes place. The data from primary control devices (i.e., RFID) are transmitted, accumulated and evaluated within the information systems level. At the interface between the information systems level and the SCEM level, SC monitoring and adaptation take place. This results in decisions on SC processes, plans or goal correcting, amending or replacing on the basis of the disturbances that occurred and the control actions that existed.

In a simple case, there is enough to establish links between RFID readers and a superordinate IT, e.g., an SCEM system. In more complex cases, the networking of different RFID readers via special protocols (e.g., low-level reader protocol (LLRP) is mandatory. In this setting, the RFID infrastructure should be built up of two levels: networking RFID readers with each other and networking RFID readers with an SCEM system.

We note that the RFID experimental environment is not intended (at least, in its current version) to the full implementation of the developed models. It is much simpler as the modeling framework and serves to gather experimental data for the modeling complex. The modeling complex itself is implemented in a special software environment, which contains a simulation and optimization "engine" of SC planning and execution control. Note that even at the execution stage we remain within the same mathematics of optimal CT and can ensure the optimization-based simulation and adaptive feedback control. The schedule execution can be analysed with regard to performance indicators and different execution scenarios with different perturbations. In the elaborated prototype, parameters of the SC structures and the environment can be tuned if the decision-maker is not satisfied with the actual schedule execution and resulting values of performance indicators. In analyzing the impact of the scale and location of the adaptation steps on the SC performance, it becomes possible to justify methodically the requirements for the RFID functionalities, the stages of an SC for the RFID elements locations, and the processing information.

In particular, possible discrepancies between actual needs for wireless solution of SC control problems and the total costs of ownership regarding RFID can be analyzed. In addition, processing information from RFID can be subordinated to different management and operation decision-making levels (according to the developed multi-loop adaptation framework). Pilot RFID devices with reconfigurable functional structure have been developed.

Let us discuss some observations and results from the experimentation stage. At the first stage of establishing the experimental stand, we came to the conclusion that traditional networking technologies tend to be expensive for integration of many readers. In this work some efforts have been made to analyze and optimize RFID network structure to be more suitable for SC control tasks. The most valuable property of RFID technology is wireless data interchange with objects (tags) without power source (passive RFID). Main advantages of passive tags such as low cost, size, weight, make them ideal candidates to be agents for data accumulation and transmission along with physical objects. Most tag access protocols are international standards. This improves interoperability of readers and tags of different manufacturers. Cost of a tag is continuously decreasing which enables their applications in real life.

Several changes can be proposed to achieve desired properties of the SC control system. First, readers should be integrated in a special bus that will provide both power and data path to readers using the single coaxial cable to be easily mounted and minimize connections. Second, software can be split according to client server model based on TCP/IP sockets using well-defined portable interface. Thus, application programmers can use the same command set to access any reader in a bus behind the server. Since the TCP/IP is an Internet protocol user applications can be spread over the network and talk to both local and remote servers.

The Sim-Sim architecture can be treated as a basement of distributed RFID networks which will be the future of RFID technology. Once the readers get inexpensive, other factors start to determine overall system cost. Sim-Sim architecture potentially offers both hardware and software solutions to reduce these costs. Existing readers can be easily integrated into Sim-Sim Server using back-end drivers. Thus, Sim-Sim architecture provides both reader manufacturers and application programmers with cost-effective, stable, scalable and portable RFID solution.

2.5 CONCLUSION

The impact of IS on the business processes in SCs becomes more and more crucial. Recent research indicated that an aligning of business processes and IS may potentially provide new quality of decision-making support and an increased SC performance. That is why it becomes a timely and crucial topic to consider SCs as collaborative cyber-physical systems. Such SCs are common not only in manufacturing but also in different cyber-physical systems, e.g., in networks of emergency response units, energy supply, city traffic control, and security control systems.

This chapter introduced the structure dynamics control concept and a dynamic model to coordinate activities in service SCs. Continuous nature of services in the SC makes them similar to continuous production systems. That is why we discussed in this chapter previous applications of CT to SCs with continuous flows. Subsequently, structure dynamics control approach was presented. Dynamic coordination information services, information resources, and business process activities have been considered on the scheduling level modeled as an OPC problem.

The proposed service-oriented description makes it possible to coordinate availability of information services and business process schedules simultaneously. It also becomes possible to determine the volume of information services needed for physical supply processes. In addition, impact of disruptions in information services on the schedule execution in the physical structure is analysed.

The results provide a base for information service scheduling according to actual physical process execution. In addition to the existing models on the scheduling of material processes in SNs, this study has added models for integrated IS, IR, and IR modernization scheduling. This study is among the first to explicitly formulate and solve in a dynamic manner the stated integrated scheduling problem. The proposed service-oriented concept allows explicitly in-corporate material and information processes in the SN and take into account modern trends of decentralized IS, e.g., cloud computing. In addition to the scheduling the proposed approach makes it possible simultaneously to (i) determine the volume of information services needed for physical supply processes (Eqs. (2.10) and (2.11)) and (ii) determine this volume in monetary form (Eq. (2.12)).

Further analysis may include an explicit incorporation of reconfiguration processes and stability into the scheduling model. In addition, the development of specialized solvers where CT and discrete optimization can be combined is a practical need to enhance the CT applications to service SC scheduling and coordination.

Table of abbreviations

CT – Control Theory
EDI – Electronic Data Interchange
EMS – Electronic Manufacturing Service
IR – Information Resource
IS – Information Service
MPC – Model Predictive Control
ODM – Original Design Manufacturer
OPC – Optimal Program Control
SC – Supply Chain
SCM – Supply Chain Management

REFERENCES

Bellmann, R. (1972). *Adaptive control processes: a guided tour*. Princeton Univ. Press: Princeton, New Jersey.

Camarinha-Matos, L.M. & Macedo, P. (2010) A conceptual model of value systems in collaborative networks. *Journal of Intelligent Manufacturing*, 21(3), 287–299.

Cannella, S., Framinan, J.M. & Barbosa-Póvoa, A. (2013) An IT-enabled supply chain model: a simulation study. *International Journal of Systems Science*, 45(11), 2327–2341.

Chen, X., I.E. Grossmann & L. Zheng (2012) A comparative study of continuous-time modeling for scheduling of crude oil operations. *Computers and Chemical Engineering*, 44, 141–167.

Chibber, P.K. & Majumdar, S.K. (1999) Foreign ownership and profitability: Property rights, control, and the performance of firms in Indian industry. *Journal of Law & Economics*, 42(1), 209–238.

Choi, T.M. (2011) Coordination and Risk Analysis of VMI Supply Chains with RFID Technology. *IEEE Transactions on Industrial Informatics*, 7, 497–504.

Choi, J., Kim, Y., Park, Y.T. & Kang, S.H. (2002) Agent-based product-support logistics system using XML and RDF. *International Journal of Systems Science*, 33(6), 467–484.

Dedrick, J., Xu, S. & Zhu, K. (2008) How does information technology shape supply-chain structure? Evidence on the number of suppliers. *Journal of Manufacturing Information Systems*, 25(2), 41–72.

Harjunkoski, I., Maravelias C.T., Bongers P., Castro P.M., Engell, S., Grossmann, I.E., Hooker J., Méndez, C., Sand G. & Wassick J. (2014) Scope for industrial applications of production scheduling models and solution methods. *Computers and Chemical Engineering*, 62, 161–193.

Ivanov D., Dolgui A. & Sokolov B. (2013) Multi-disciplinary analysis of interfaces "Supply Chain Event Management – RFID – Control Theory". *International Journal of Integrated Supply Management*, 8, 52–66.

Ivanov, D. & Sokolov B. (2012a) Dynamic supply chain scheduling. *Journal of Scheduling*, 15(2), 201–216.

Ivanov D. & Sokolov B. (2012b) Structure dynamics control approach to supply chain planning and adaptation. *International Journal of Production Research*, 50(21), 6133–6149.

Ivanov D. & Sokolov B. (2013) Control and system-theoretic identification of the supply chain dynamics domain for planning, analysis, and adaptation of performance under uncertainty. *European Journal of Operational Research*, 224(2), 313–323.

Ivanov D., Sokolov B. & Dilou Raguinia, E.A. (2014) Integrated dynamic scheduling of material flows and distributed information services in collaborative cyber-physical supply networks. *International Journal of Systems Science: Operations & Logistics*, 1(1), 18–26.

Ivanov D., Sokolov B. & Dolgui, A. (2013) Multi-stage supply chain scheduling in petrochemistry with non-preemptive operations and execution control. *International Journal of Production Research*, 52(13), 4059–4077.

Ivanov, D., Sokolov B. & Dolgui, A. (2012) Applicability of optimal control theory to adaptive supply chain planning and scheduling. *Annual Reviews in Control*, 36, 73–84.

Ivanov, D., Sokolov, B., Dolgui, A., Werner, F. & Ivanova, M. (2015). A dynamic model and an algorithm for short-term supply chain scheduling in the smart factory Industry 4.0. *International Journal of Production Research*, in press.

Jain, V., Wadhwa, S. & Deshmukh, S.G. (2009). Revisiting information systems to support a dynamic supply chain: Issues and perspectives. *Production Planning and Control*, 20(1), 17–29.

Jiang, N., Xu, L., Vrieze, P.D., Lim, M.G. & Jarabo, O. (2012) A Cloud Based Data Integration Framework'. In: Camarinha-Matos, L., Xu, L. & Afsarmanesh, H. (eds.) *Proceedings of the IFIP Conference on Virtual Enterprises PRO-VE 2012 IFIP AICT 380*, Springer, pp. 177–185.

Jula, P. & Kones, I. (2013) Continuous-time algorithms for scheduling a single machine with sequence-dependent setup times and time window constraints in coordinated chains. *International Journal of Production Research*. 51(12), 3654–3670.

Perea, E., Grossmann, I., Ydstie, E. & Tahmassebi, T. (2000) Dynamic modeling and classical control theory for supply chain management. *Computers and Chemical Engineering*, 24, 1143–1149.

Pershke E. (2012). *How to Create a World-Class Supply Chain: IndustryWeek* [Online] Available from: http://www.industryweek.com/supply-chain/how-create-world-class-supply-chain?page [Accessed 29th July 2015].

Pontryagin, L.S., Boltyanskiy, V.G., Gamkrelidze, R.V. & Mishchenko, E.F. (1964) *The mathematical theory of optimal processes*. Pergamon Press, Oxford.

Puigjaner, L. & Lainez JM. (2008) Capturing dynamics in integrated supply chain management. *Computers and Chemical Engineering*, 32, 2582–2605.

Sarimveis, H., Patrinos, P., Tarantilis, C.D. & Kiranoudis, C.T. (2008). Dynamic modeling and control of supply chain systems: A review. *Computers & Operations Research*, 35, 3530–3561.

Sethi, S.P. & Thompson, G.L. (2006). *Optimal Control Theory: Applications to Management Science and Economics*. Second Edition, Springer, Berlin.

Shah, N. (2004). Process industry supply chains: Advances and challenges. *Computer Aided Chemical Engineering*, 18, 123–138.

Subramanian, K., Rawlings, J.B., Maravelias, C.T., Flores-Cerrillo, J. & Megan L. (2013) Integration of control theory and scheduling methods for supply chain management. *Computers & Chemical Engineering*, 51, 4–20.

Wang L.C., Cheng C.-Y. & S.-K. Lin (2013) Distributed feedback control algorithm in an auction-based manufacturing planning and control system. *International Journal of Production Research*. 51(9), 2667–2679.

Wang, L., von Laszewski, G., Younge, A., He, X., Kunze, M., Tao, J. & Cheng, F. (2010) Cloud Computing: a Perspective Study. *New Generation Computing*, 28(2), 137–146.

Zhuge, H. (2011) Semantic linking through spaces for cyber-physical-socio intelligence: A methodology. *Artificial Intelligence*, 175(5–6), 988–1019.

Chapter 3

Signs in service supply[1]

Helge Löbler
Institute for Service and Relationship Management ISRM, Universität Leipzig, Leipzig, Germany

SUMMARY

The world is full of signs (symbols, signifiers, words included). They guide us in supermarkets, on highways and in airports. They even guide us to the right bathroom. Signs also help us when we read manuals or when we use a remote control. They are taken for reasoning important decisions in the medical as well as in the financial and business world. Signs are used everywhere. A less understood characteristic of signs is that they render service in very different ways. This chapter analyses how signs render service when "carrying" meaning, when they imply actions and when they change thinking. The findings reveal how signs affect services which are important to support future research on signs related service supply chain management.

Keywords
Signs, signifiers, practices, service, resources, meaning and innovation

3.1 SIGNS RENDER SERVICE, WHEN THEY IMPLY MEANING

3.1.1 Signs are ubiquitous

Signs are omnipresent. They guide us when we are shopping in supermarkets, on highways and in airports. They even guide us to the right bathroom. Signs have a huge impact if they are used as logos or in advertisements. Signs also help us when we read manuals or when we use a remote control. Signs are used everywhere. They give us orientation to find what we are looking for, whether we are looking for a particular product, service or any kind of item or activity or event.

In addition, in many service situations customers are not interacting directly face-to-face with other people but with technical devices or technologies. This holds for

---

[1]This chapter is based on several former articles: Löbler 2010; Löbler 2014; Löbler and Lusch 2014 and Löbler and Wloka 2015. It generalizes the findings described therein and makes them available in a broader perspective for an easier use.

commonly used teller machines, for online-banking, interactive kiosks, interactive voice responses, and artificial virtual agents, etc. Not to forget the million apps used on mobile devices or the manuals, which we read over and over, trying to understand them. When one does not directly interact with other people face-to-face the "interaction" is very often reduced to reading and writing, i.e. to signs and their "understanding". Supply management has to be aware of the signs which accompany the whole supply chain process as signs are important resources and governing instruments. In addition feedback loop in supply chain management is mainly represented by signs.

I will use the term "sign" in the following sense: a sign is something that is perceivable as a sign (not noise), something we become aware of through the senses. A sign is perceived as a sign as it is related to something else than the sign itself. I use "sign" and "signifier" synonymously and also to denote arrangements of signs or signifiers, such as words or sentences (Löbler 2010, p. 20; Löbler and Lusch 2014). In a world of sensory or multisensory marketing signs are probably not limited to the senses of hearing or seeing. An extended understanding of signs in the contemporary world could integrate at least the other traditionally recognized senses. Together with the two mentioned they are: Sight (ophthalmoception), hearing (audioception), taste (gustaoception), smell (olfacoception or olfacception), and touch (tactioception). In this understanding we don't only have audio or visual signs but also gustatory, olfactory and tactile signs. Singapore airlines, for example, has created a specific fragrance, Stefan Floridian Waters. This patented aroma has been specifically designed to complement the airline's brand, and has become their trademark fragrance. The smell now is related to Singapore Airlines. One important distinction of signs in the above definition is that only visual and audio signs can be digitized. The reason is that taste, smell and touch cannot really be separated from matter and one can doubt that these three senses can be used to define signs in the above sense.

3.1.2 Characteristics of signs

In modern cities there are millions of signs around us. If one would read and understand all of these signs one would become crazy. We as human beings have to ignore most of the signs around us in modern cities. This is a selection of those signs being of some importance. The signs which are ignored don't render service to those ignoring them. They may render service to other people ignoring other signs. The ubiquity of signs is not limited to visual signs or things which are seeable. As mentioned it also holds for acoustic signs. It is interesting that the evolution of language has generated a specific term for all the acoustic signs we ignore or don't want: noise. It is the sound we don't want to hear. Merriam Webster defines noise as "any sound that is undesired or interferes with one's hearing of something". A similar term is now emerging in IT: "Visual Noise". However it is defined very differently depending on the field of research. Engberg for example defines "the term 'visual noise' as a distinctly definable strategy which combines letters with images, sounds, and, in the case of digital work, kinetic operations to create a sense of excess" (Engberg 2010). In product design Baskinger defines visual noise as the discrepancy between what the user sees and what the user understands. (Baskinger n.d.).

As a general definition I propose: visual noises are any seeable signs that are undesired or interfere with one's seeing of something. The main characteristic of visual noise is that it is seeable and unwanted. This short discussion about visual and acoustic signs and noise shows clearly that signs do not render service just because they are signs. If they are not wanted they are not service but noise. Using of signs is thereby a functional support of a state or activity whereas a visual noise has a dysfunctional potential or impact to a state or activity.

A main characteristic distinguishing noise from signs is that a sign draws a distinction whereas noise does not. To be precise noise does not draw a distinction between itself and other noise whereas a sign draws a clear distinction between itself and other signs. A sign also draws a distinction between noise and itself.

3.1.3   Signs and their referent

Signs and their drawing a distinction is rooted in Spencer-Brown's (1969) calculus of form, which states that whenever one names something, the precondition of marking it by a name is a distinction. The drawing of the distinction and the marking of one side of the distinction is always done simultaneously. Calling something a "tree" or "car" not only names it but also distinguishes it from all other things that are not a tree or a car. Thus, naming something a "tree" or "car" makes a distinction. This distinction creates a difference, a difference by which Bateson defines information. I explicitly exclude Shannon's (1998) perspective here because it neither involves meaning nor refers to entities of reference to which we come soon. Shannon notes that "semantic aspects of communication are irrelevant to the engineering problem" (p. 31). Weaver (1998, p. 8) interprets this sentence as follows: "In fact, two messages, one of which is heavily loaded with meaning and the other of which is pure nonsense, can be exactly equivalent, from the present viewpoint, as regards information." In contrast, Bateson (1970, p. 7) defines information as "the difference which makes a difference." This definition supports our analysis (see also Löbler 2010, p. 20; Löbler and Lusch 2014).

What kind of difference is Bateson talking about? Consider the following introductory example. English-speaking Person A says to English-speaking person B, "Yesterday a "nuhai"[2] came home." Person A makes a difference by distinguishing a "nuhai" from the rest of the world. However, this difference does not necessarily make a (second-tier) difference for person B if person B has no idea what or who a "nuhai" is. If, however, A says, "A girl came home yesterday," person A makes a first-tier difference by distinguishing the girl from the rest of the world, and this distinction makes another, second-tier difference for person B because now the word "girl" refers to a human being, not just a word. For person B, the word "nuhai" did not refer to a girl, which could be distinguished from other entities in the world; therefore, it did not make a second-tier distinction or difference. In short, using a word makes the first-tier distinction, and if the word refers to an entity that can be distinguished from the rest of the world (second-tier distinction), the word becomes an information in the Bateson sense. However a sign becoming an information does not necessarily assure understanding. Take the simple sentence "I did a regression" as an example. The word regression is a distinction/difference that makes a difference. However it can make a different difference for different people. For a statistician the word "regression"

---

[2]Nuhai "Nühái" is the Chinese word for girl.

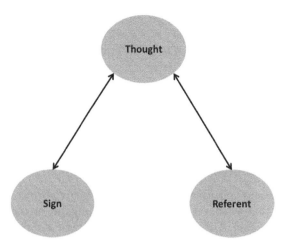

Figure 3.1 Relation of sign, thought and referent. The Meaning of Meaning. A study of the influence of language upon thought and of the science of symbolism from Ogden and Richards, 1946.

has a fundamental different meaning as for a psychotherapist. According to Merriam Webster for a psychotherapist it is "reversion to an earlier mental or behavioral level" whereas for a statistician it is "a functional relationship between two or more correlated variables that is often empirically determined from data and is used especially to predict values of one variable when given values of the others". Hence we have to introduce a third tier difference to cover not only information and individual understanding but also to cover understanding and misunderstanding between people. The first tier difference is the word itself as being different from other words. The second tier difference is the idea or cognitive representation associated to the first tier difference. It is what we think in association with a word or sign. However what person A thinks when hearing a word is not necessarily the same to what person B might think hearing the same word.

The third tier distinction is a referent to which the word and the cognitive representations (thought) refer to. It is important to be aware that "between the symbol and the referent there is no relevant relation other than the indirect one, which consists in its being used by someone to stand for a referent. Symbol and referent, that is to say, are not connected directly." (Ogden and Richards 1946, p. 11) (see figure 3.1). In the light of social constructionism we can call the referent a socially construed representation in the following sense: The "society" in which one uses a word or sign relate the referent to that word. In a society of therapists the word regression is related to the process or activity of reversion to an earlier mental or behavioral level. In the society of statisticians the word regression is related to the process or activity of making a functional relationship between two or more correlated variables that is often empirically determined from data and is used especially to predict values of one variable when given values of the others.

As Löbler explains that "the postmodern perspective—especially the poststructuralist perspective—goes even further in saying that there is not even an indirect relationship between the sign and the referent. Baudrillard (1975, p. 128) and Derrida (1976, 1977, 1978) totally disconnected the sign as a signifier from the referent: "The sign no longer designates anything at all. It approaches its true structural limit which is to refer back only to other signs". From this perspective Cherrier and Murray (2004, p. 513) conclude "In the post-modern era, there is no longer an attempt to refer back to nature or ground the representamen." This perspective is used by Venkatesh et al. (2006, p. 251) in their emphasis on "[...] (re)considering the starting point of our disciplinary analysis to be the market [...] as opposed to marketing [...]," where they considered the "[...] market as a sign system [...]" (2006, p. 258). But if markets and signifiers in general are "only" plain sign systems how (if at all) are these sign systems linked to activities or (inter-)actions (doings)? How are they understood to be transformed into (inter-)action?" (2010, p. 218). Everybody experiences strongly that signs and signifier are somehow connected to the real world outside of language. Hence the deconstructivist perspective leaves a dead-end road for our purpose.

Instead it is proposed here that signs render service in different ways when they are understood as complimentary to practices. Practices are ways of doing not the doing itself. It is the social representation of how activities are performed or should be performed. In performing a particular practice, one may notice a pattern and then may name the pattern. For example, we say "googling" or "tweeting" in English or "googeln" or "twittern" in German for these new practices.

In his Philosophical Investigations Wittgenstein (2008, §199) asks what it means "to obey a rule". He explains that *thinking* one is obeying a rule, *saying* one is obeying a rule and *obeying* a rule is not the same. "Obeying a rule is a practice" (Wittgenstein 2008, §202). From Wittgenstein's work there is a direct path to Schatzki's (1996) "Practice Theory".1 "One of the watchwords in contemporary humanistic thought about human activity is 'practice.' (Schatzki 2007, p. 98).

For most theorists, meanwhile, the term serves as a signal that such phenomena as identity, language, gender, science, and social organization, which had not been previously construed in the following way, are best thought of as **rooted in or as forms of activity**. (Schatzki 2007). For Giddens, too, practices are the principal unit of investigation: "The basic domain of study of the social sciences, according to the theory of structuration, is neither the experience of the individual actor, nor the existence of any form of societal totality, but social practices ordered across space and time." (Giddens 1984, p. 2).

Explicit coordinations are embedded in a lifeworld background (lebensweltlicher Hintergrund) to use Habermas's term (1985a, b) and "[...] in general, cultural theories (and especially theory of practices) relativize the rationalist models of the interest-following or the norm following transparent agent by situating action in implicit or unconscious, collective symbolic structures." (Reckwitz 2002, p. 261) In particular, "Practice theory—as it is exemplified in authors such as Bourdieu, Giddens, late Foucault, Garfinkel, Latour, Taylor or Schatzki—is a type of cultural theory." (Reckwitz 2002, p. 245) Therefore practices may serve as a layer for implicit coordination (Espinosa et al. 2004, Toups and Kerne 2007, Löbler 2010).

"A practice is thus a routinized way in which bodies are moved, objects are handled, subjects are treated, things are described and the world is understood." (Reckwitz, 2002, p. 250) "This way of understanding is largely implicit [...]" (Reckwitz 2002, p. 249). The theory of practices has been discussed in marketing, especially in a special issue of Marketing Theory in 2008, and there are attempts to integrate it into a broader service systems perspective (Vargo and Lusch 2011). As a general theory a "[...] practice approach stands in opposition to individualist ontologies where social phenomena are viewed as products arising out of the actions and mental states of individuals, and societism understood as the study of social facts, structures and systems that resist reduction to individual actors." (Araujo et al. 2008, p. 6). According to Warde (2005, p. 147) it is in opposition to a sign/signifier-oriented stream, as he makes clear: "Theories of practice also provide a powerful counterpoint to expressivist accounts of consumption." Simultaneously, Schatzki's notion of practices as a "nexus of doings and sayings" (Schatzki 1996, p. 89) builds an underlying, connecting bridge between expressed sayings and signs on the one hand and doings on the other hand. Schatzki (1996), like Habermas (1985a, 1985b), addresses the implicitness of practices (Schatzki) and the lifeworld background (lebensweltlicher hintergrund, Habermas). In this sense practice theory is not only a nexus of doings and sayings: it also connects body and mind, and the individual subject is the "carrier of the practice" (Reckwitz 2002, p. 250). Hence it serves as an implicit layer in which all explicit coordination is embedded. Practices coordinate ways of doings, sayings and thinking, and concomitant practices are created by different ways of doing, sayings and thinking. So we see the theory of practices as the link not only between our realms, but also between the different focuses of these realms. Practices are implicitly behind all forms of explicit coordination, they coordinate implicitly, and we can become aware of them by the ways we do or say things (Löbler 2010). "These practices, overhearing and ambient monitoring, aid in implicit coordination capabilities." (Toups and Kerne 2007, p. 714).

From a business perspective, customer practices cannot be understood by simply looking at the signs used. Signs indicate different practices for different people. Firms must also understand or experience customer practices, especially if they want to support or offer substitutes. For instance, when surgeons switch from traditional to robotic assisted surgery, it is important to understand the surgeons' experiences in such things as how to move their wrists and then to use this information to design better training programs, usually with computer simulation, for the surgeons.

Löbler and Lusch (2014) explain that service often integrates practices by the use of signs, and by doing so, practices become resources, something an actor draws on for support (von Krogh et al. 2012). Grönroos (2008) and Sheth et al. (2000) mention practices from the perspective of service-dominant logic, but they do not use the term in the same vein as Schatzki (1996) and Reckwitz (2002), who follow Wittgenstein's (1960) understanding of practices as implicit ways of doing. Rather, they use the term as a synonym of doing. Reckwitz emphasizes the different meanings using the German words "praktiken" and "praxis" (as a synonym for doing). Note that the term "practices" ("praktiken") as understood in practice theory has a very different meaning than practices in the vein of praxis, which is simply what is done. Löbler's (2010) analysis of signs and practices (also in accordance with service-dominant logic) follows the meaning of practices in the sense of praktiken. Similarly, in their study of how brand community practices collectively create value, and consistent with value cocreation in service-dominant logic, Schau et al. (2009) adopt the perspective along a praktiken concept of practices. Thus, it is not just simply what people do, but it is a theoretical approach to practice that stands in contrast to individualist ontologies and where social phenomena are understood as mainly societal facts, structures, and systems that resist reduction to individual actors (Araujo et al. 2008, Löbler 2011).

3.1.4 Signs and practices working together

It is helpful to distinguish between thoughts as individual representations of a referent and practices as a social representation of a referent. Signs connect both. Second order cybernetics is offering valuable concepts for relating signs, thoughts and referent. These concepts are particularly reclusive operations and eigenforms.

We propose that meaning can be understood as a stable form of recursive operations in language use, so called eigenforms (Löbler and Wloka 2015). These recursive operations are typically chains of performed practices forming a loop together. Eigenforms serve as meaning giving stable process characteristics of chains of performed practices. We use the term and concept of eigenform as it was put forward by von Forester (1981) and further developed and used by Kaufmann (2003, 2005) or Füllsack (2012).

Second order cybernetics is used to explain how objects appear as eigenvalues or "token of eigen-behaviour" in recursive operations (von Foerster 1981). Von Foerster proposes that objects are tokens of eigen-behavior and that they created by a recursive operation between observing and acting or more precisely the coordination of action (2003). These ongoing recursive operation may reach a stable state as an eigenform which according to von Foerster is ontologically not distinguishable from an object (see figure 3.2). Hence the eigenform appears as an object or an eigenform out of the recursive operation of observing and acting.

"Ontologically, eigenvalues and objects, and likewise, ontogenetically, stable behavior and the manifestation of a subjects "grasp" of an object cannot be distinguished." In both cases "objects" appear to reside exclusively in the subject's own experience of his sensori-motor coordinations; that is, "objects" appear to be exclusively subjective!" (Forester, 2003, p. 267). This process is in line with the direct relation between thoughts and referent.

In the same vein Löbler and Wloka (2015) have used second order cybernetics to understand meaning as an eigenform of recursive operations.

In the social realm the object and the indication of an object (creating a sign, symbol or word) emerge simultaneously. We have seen that for von Foerster objects are "created" by a subject (individual) though a recursive operation of observation and

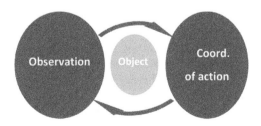

Figure 3.2 Eigenform of the recursive operation. Understanding understanding: Essays on Cybernetics and Cognition from von Foerster, 2003.

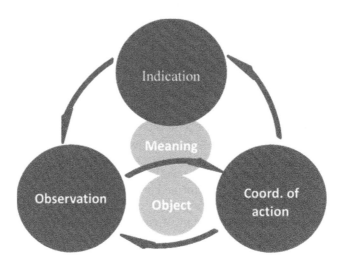

Figure 3.3 Recursive process for an individual.

activity as shown in figure 3.2. If more than one subjects come into play they have the individual observation and activity recursions and in addition a recursive operation with the other individual by use of indication. These indications are now part of the recursive process as shown in figure 3.3 and for two individuals in figure 3.4.

As an object appears as a stable form of a recursive operation now the meaning appears as a stable form of indicating, observing and acting. And furthermore if the indication is an abstract term like "loyalty" (Löbler and Wloka 2015) the object and the meaning coincide. This is exactly the indirect relation of sign and referent Odgen and Richard talked about (Ogden and Richards 1946). There is no "real" reference to an abstract term. Hence if objects are an eigenform of recursive operations and if indication is part of this recursive operation then words refer either to other words or to activities. This holds for abstract term in particular. Hence our main propositions are. Löbler and Wloka come to the conclusion that "words refer either to other words or to activities (and their observations)" and that "meaning is an imaginary eigenform which can be used as being real." (Löbler and Wloka 2015). As a simulation of language use Löbler and Wloka used the Merriam Webster dictionary and were able to show how stable states (infinite loops) of recursive operations emerge for the word "loyalty". They identified two forms of stable states: 1) Closed loops between words only and 2) closed loops between doing and indicating identified as descriptions of activities (see figure 3.5). Their example additionally showed that meaning can be understood as an eigenform of recursive operations. Whereas von Foerster's operations created objects as token of eigen-behaviour, indication is needed in the social realm as part of the recursive operations to create meaning as an eigenform. Eigenforms of observing and coordinating action lead to individual representation inaccessible for others. Eigenfoms of communicating signs and coordinating actions between individuals are practices. The latter appear as meaning.

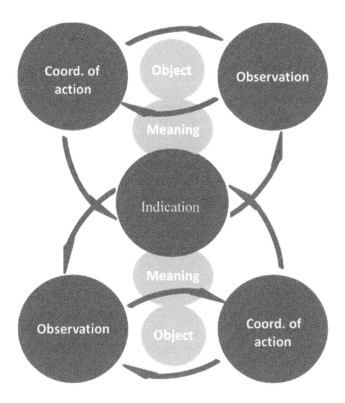

Figure 3.4 Recursive process for two individuals.

Most of us intellectually believe in the following statement while some of us may not. "The meaning of a word is its use in the language" (Wittgenstein 2008, § 43). By using words and with them language it seems that these words have meaning or rather that they are used in such a way that they appear to have meaning. In some social situations people even fight for the appropriateness of the meaning of the words they use (for instance when applying law). We use words as if they would mark real entities. However we have given an interpretation of what Wittgenstein indicates as "its use in language". When words don't have meaning in themselves nobody can claim the right use of words. Different reclusive operations may lead to different meanings. And different recursive operations appear in different social communities. To be aware that signs have to meet different eifgenforms (meanings) if spread out into different social communities is one of the most important conclusion of the above considerations. Signs only render service if they are understood, if they find an eigenform in the social community used. Simultaneously inside a social community specific signs have to be used when coordinating specific processes. In this sense experts like medical doctors or lawyers have their jargons which are usually not understandable for lay actors. However if these Experts miss their own jargon it can lead to fatal consequences.

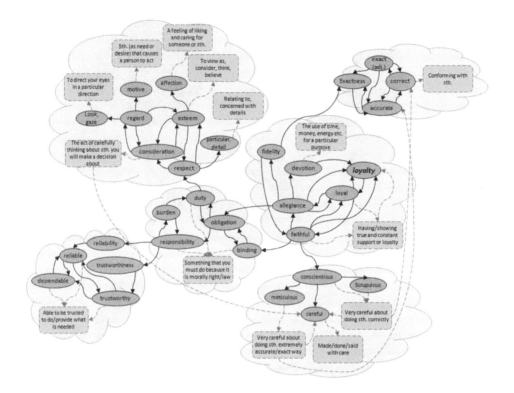

Figure 3.5 Stable states (infinite loops) of recursive operations. Loyalty' Between Talk and Action – Meaning as Eigenforms of Recursive Operations from Löbler and Wloka, 2015.

3.2   SIGNS RENDER SERVICE, WHEN THEY IMPLY ACTIVITIES

The following example show what can happen if experts are talking outside their jargon, a case in which the absence of specific signs is extremely harmful: On January 25, 1990, a Boing 707 crashed at 21:34 approximately 20 miles before landing at John F. Kennedy International Airport. Eight of the nine crew members and 65 of the 149 passengers on board were killed. The pilots of flight Avianca 052 were not able to communicate their problems to the tower. They simply didn't know or use the right words (signs) to communicate their extremely dangerous situation to the air traffic controller. This word is: Emergency. Since the pilots did not use it the air traffic controller could not know how serious the situation of the aircraft was. One word could have saved 73 lives (NATIONAL TRANSPORTATION' SAFETY BOARD WASHINGTON, D.C. 20594 1990). This is by no means a single case. Cushing (1995) and Barker (2012) describes more cases of fatal miscommunication in aviation. Typical for these cases is that pilots or traffic controller (mostly pilots) did not use the appropriate jargon. However for an expert controller only the right jargon is connected to the relevant practices and actions to be done. If the signs given by e.g. a pilot do not resonate to the relevant practice which is the expert's referent they don't cause the relevant actions.

Another fatal miscommunication is the case in which an expert jargon is used misleadingly and receivers of this communication are not aware of the misleading communication. This kind of inappropriate communication caused the US financial crisis in 2008 (Löbler 2014). Rating agencies used high ratings for bad investments. In the former case the signifier 'emergency' was the right one to be used to signify the situation in the latter the signifier AAA was often the wrong to be used. We all know that misunderstandings happen and misunderstandings are common situations for most of us. However we run into an enormous risk if we don't know, if we are not aware, that we are a part of a misunderstanding. The pilots of flight Avianca 052 didn't know that they hadn't communicated their problems appropriately. Many investors didn't know that the ratings didn't indicate what they should indicate. How can this happen?

According to Giddens (Giddens 1991b, 1991a) and Luhmann (Luhmann 1995, 1996) societal systems can be distinguished into face-to-face systems and faceless commitments. Giddens (1991a) notes that "the former refers to trust relations which are sustained by or expressed in social connections established in circumstances of co-presence. The second concerns the development of faith in symbolic tokens or expert systems, which, taken together, I shall term *abstract systems*" (p. 80, emphasis in original). Giddens (1991a) defines expert systems as "systems of technical accomplishment or professional expertise that organize large areas of the material and social environment in which we live today." (p. 27). Modern life is full of expert systems and their jargons. Most houses involve a network of expert systems: consumers rely on the heating system as well as, the energy flow. When traveling, consumers rely on the airplanes as well as on the pilots' ability to fly them. In addition, most people are part of expert systems; attorneys, architects, physicians, and rating agencies can be understood as expert systems in this sense. Expert systems emerge according to the division of labor and specialized skills or knowledge (Löbler 2014). Thus, they are part of the service society.

In addition expert systems need to communicate. There is not only communication inside an expert system but also between different expert systems and between expert systems and laypeople. In these cases symbolic token do not really provide a guarantee of expectations across 'distanciated' time-space. They are taken away from their area of use in which they have clear meaning. They are using Giddens's terminology 'disembedded'. Disembedding means "the 'lifting out' of social relations form local contexts of interaction and their restructuring across indefinite spans of time-space" (Giddens 1991b, p. 21). Löbler (2014) as shown how this disembedding of symbolic token has influenced the US financial crisis (see below). In such cases signs do not provide service but to the contrary they have harmful potential. But how can one know what the symbolic token means if one does not belong to the expert system in question? In face-to-face situations one can simply ask the person using a symbolic token what she means by using it. However in faceless commitments when communicating with expert systems indirectly there is usually no chance to ask. This does not only mean that one can misunderstand a symbolic token but also that one may not even be aware of a misunderstanding. Gadamer pointed this out clearly: "How can one be aware of a misunderstanding when reading a text (here: symbolic token), if the text does not contradict?" (Gadamer 1990, p. 273). Hence disembedded symbolic token can not only be misunderstood but in addition the misunderstanding may not become aware,

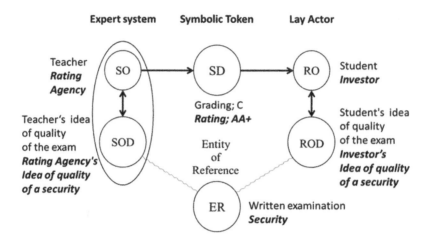

Figure 3.6 Expert systems and functioning of symbolic token. When Trust Makes It Worse—Rating Agencies as Disembedded Service Systems in the U.S. Financial Crisis from Löbler, 2014.

it may not become explicit, it remains unnoticed unless there is some contradiction to the expectation. Figure 3.6 is a visualization of expert systems and the functioning of symbolic token (taken from Löbler 2014).

In a simple sender–receiver model, one can distinguish one first-tier difference, the word or sign used; two second-tier differences, one the sender has in mind and one the receiver has in mind; and an entity of reference to which both second-tier distinctions refer. Figure 3.6 presents the model graphically with the following definitions:

1   the difference the sign indicates in that it is a sign (not noise) (the sign difference (SD), the first-tier difference)
2   the difference a sign sending observer (SO) wants to indicate by using the sign (intended meaning; the sending observer's difference (SOD), a second-tier difference)
3   the difference a sign-receiving observer (RO) indicates by receiving the sign and (perceived meaning; the receiving observer's difference (ROD), a second-tier difference)
4   an entity of reference (ER) (if available) to which the differences refer

Often, SOs are experts in their fields (e.g., physicians, attorneys, and teachers). These groups of people form what Giddens (1991a) calls an expert system that provides service; they are people of specific expertise that lay actors typically do not have. However, not all expert systems create symbolic tokens.

Physicians and attorneys, for example, have terminology but do not use these symbolic tokens to communicate to lay actors; they try to use the lay actors' language to communicate to their clients. In contrast, teachers as expert systems use symbolic tokens—namely, the grading system (as described previously)—to communicate to lay actors. Both the expertise and the grading system provide service. Figure 3.1 provides an overview. It connects Bateson's (1970) idea of information to a well-known

Table 3.1 Synopsis of Descriptions (Löbler 2014).

General description	Indicated by	Grading by teachers	Rating by rating agencies	Giddens's terminology
Difference a sign indicates (first-tier difference)	SD	C (A, B, C, D, E, F scale)	AAA (AA+, A A etc., scale)	*Symbolic token* if the signs used are independent from any reference
Sending observer	SO	Teacher	Rating agency	Expert as part of an *expert system*
Receiving observer	RO	Student	Investors	
Sending observer's difference (second-tier difference)	SOD	Teacher's idea of quality of the exam	Rating agency's idea of the quality of a security	Expertise as part of an *expert system*
Receiving observer's difference (second-tier difference)	ROD	Students' idea of quality of the exam	Investors' idea of the quality of a security	
Entity of reference	ER	Written examination	Security	
Other observers	OO	Parents, classmates, etc.	Financial authorities, homeowners, etc.	
Other observers' difference (other second-tier differences)	OOD	Parents', classmates' and other people's idea of the quality of the exam	Financial authorities, homeowners, and other people's idea of a security	

example (a teacher's grading system), and it connects the terminology to both forms of abstract systems: an expert system and symbolic tokens. A teacher (SO) grades a written examination (ER) of a student (RO) with a C (SD). The teacher's idea of the examination's quality is his or her difference as indicated by a C (SOD). The student's idea of his or her examination is his or her difference indicated by a C (ROD). The C is the difference the sign C indicates in this context; it is a sign and not noise (SD), the teacher's idea (SOD) of the written examination causing the grading, the student's idea of the received C (ROD), and finally, the written examination as the entity of reference (ER). There might also be other observers (OO) with their own differences, such as classmates and parents and other observers' differences (not shown in Figure 3.1).

The similar reasoning holds for the rating agencies: The rating itself is the first difference (e.g., AAA, AAC; SD). The rating agency (SO) uses the sign according to its idea of quality (SOD) of a security (ER) to support investors (RO). Reading the rating, investors have an idea of the quality of the security (ROD), on which they base their investment decisions. Table 3.1 summarizes the descriptions.

As described in section 1 the entity of reference is out of reach of language or out of reach of at least one of the participating parties. In the case of flight Avianca 052 the tower control had no access to the airplane situation. In case if the US-financial crisis investors had no idea about the securities they invested in. The more abstract signs become the more unclear is the entity of reference. In the same vein expert jargons do not refer to something for lay actors. The real challenge in these kinds of situation is that ignorance appears as knowledge.

As described in Table 3.2 there are four different situations A-D between awareness and ignorance. The captain of flight Avianca 052 as well as most investors during the US-financial crises were not aware of their own ignorance.

Table 3.2 Four different situations of awareness and ignorance.

	I know	I don't know
I am aware that...	A I know the size of my shoes. I know that I know the size.	B I don't know the length of the Mississippi river. I know that I don't know the length.
I am not aware that...	C All tacit knowledge belongs to this category until it can be explicated precisely. E.g. most parts of a grammar of a language are used 'automatically' without explicitly knowing it. Or when riding a bike we are not aware how it is done.	D If one is not aware that one does not know, this ignorance appears as knowledge: In answering questions of an examination students sometimes give wrong answers, but they are not aware that their answers are wrong. They think they know. Unless there is a contradiction or an unexpected activity.

Redundancy and variety are ways to avoid that ignorance which appears as knowledge. Discussing common grounds in which parties involved discuss their views (inter-views) are helpful in discovering unawareness of ignorance.

3.3 SIGNS RENDER SERVICE, WHEN THEY IMPLY IMAGINATIONS, INVENTIONS IN PARTICULAR

Although business scholars (Berry et al. 2006, Michel et al. 2008, Ordanini and Para-suraman 2011, Sawhney et al. 2005) are finally focusing on service versus goods inventions and innovation, there is a much broader interest. This includes economics (Cainelli et al. 2006, Gallouj 2002, Gallouj and Suvana 2009), strategy (Dorner et al. 2011), and information systems (Fichman et al. 2014, Nambisan and Sawhney 2007, Sheehan 2006, Swanson 1994, Yoo et al. 2010). Importantly, some argue that service innovation is distinct from product (goods) innovation and thus new theories and frames of reference are required (Edvardsson and Olsson 1996, Fitzsimmons and Fitzsimmons 2000, Kallinikos et al. 2013). However, others argue that an integrative framework that includes services and products and thus avoids the goods–services divide is needed; these studies try to modify current innovation models to the service innovation arena (Nijssen et al. 2006). In this regard, Vargo and Lusch (2004, 2008, 2011) would argue that service-dominant logic can provide a unifying innovation framework because it views all innovation as based on the novel integration of resources to support value co-creation (Löbler and Lusch 2014).

All service innovations are created by inventions or new thoughts and then show the form of signs and practices, which become resources to provide services.

It is argued by Löbler and Lusch (2014) that the view of signs and practices as resources is underplayed in their work and mostly ignored in other work in marketing related to practices. In addition, practices have not been discussed as resources in IT-related service innovation, though these practices (in the vein of praktiken) are very important in service and IT-related service in particular because they are the counterpart of signs and serve as meaning providers of signs.

Let's illustrate this connection by an example originally from a nondigital world and now common in the IT world (Löbler and Lusch 2014). Consider the now familiar computer "desktop," which is commonly used as a user interface for interacting with the internal computer or IT world. It was not always this way. A computer desktop is kind of a simulation of a real desktop. It works as a user interface because people were familiar with desks and using their desktops to work with and arrange their papers and other items. Using a desk and its desktop was a practice long before computers. The first personal computers did not even have desktops, with user interaction done solely through symbols (signs) on a command line. Apple introduced the desktop to personal computing in the 1980s, adapting a method pioneered by Xerox for business computing (Smith et al. 1982; see also Erickson 1990). People could suddenly use existing practices (working on a desktop) as resources when working with a computer. The computer desktop is simulated by signs, which can be embedded in real-world performed practices. So both the signs and the practices become resources. In this case, people can intuitively use IT because its use is based on practices. The intuitive uses are always practice based but are not always explicit. If practices are integrated, they become resources. Moving documents on a touch screen feels almost identical to moving items on a real desktop—just touch the document icon and move it using an implicit practice that we are all used to, but not always aware of. These kinds of innovations integrate practices and signs: the practice becomes operative by moving fingers like touching a document and then drawing it up or down. The signs (on the screen) create or simulate the impression that the documents are really moving.

The distinction between signs and practices allows describing paths of, e.g., IT-related innovations as well as finding new relevant areas for further innovations (Löbler and Lusch 2014).

A new practice was, for example, generated with the invention of mechanical typewriters and hence a new market. Typewriting was simply not existent to write on a tablet's screen with a pen-like device as an integration of an old practice (handwriting) into a new technology and thus can be viewed as dynamically continuous (i.e., there is some, but not a major, alteration of usage and behavior patterns). Figure 3.7 shows the pathways to IT-related innovation, using "writing" as a part of "word processing," along two independent dimensions: sign operations that are enabled by technology and practices that are integrated when these technologies are used. (Löbler and Lusch 2014).

Writing is a well-known practice performed by using materials such as paper and an instrument such as a pencil as resources operating with signs. There were some innovations throughout history, such as different types of ballpoint pens and fountain pens; however, these innovations did not change the practice of writing (indicated by arrow 1 in Figure 3.7) but made it easier. The invention of the mechanical typewriter instead changed the practice of writing (indicated by arrow 2 in Figure 3.7). People had to be well-trained to use mechanical typewriters. To exercise this practice was part of secretaries' education. Then electronic typewriters were introduced (indicated by arrow 3 in Figure 3.7). It did not change the practice of typewriting in principle, but it required the typist to learn the sensitivity of the keys on the typewriter so as to type faster and more effectively. Then another innovation was the introduction of

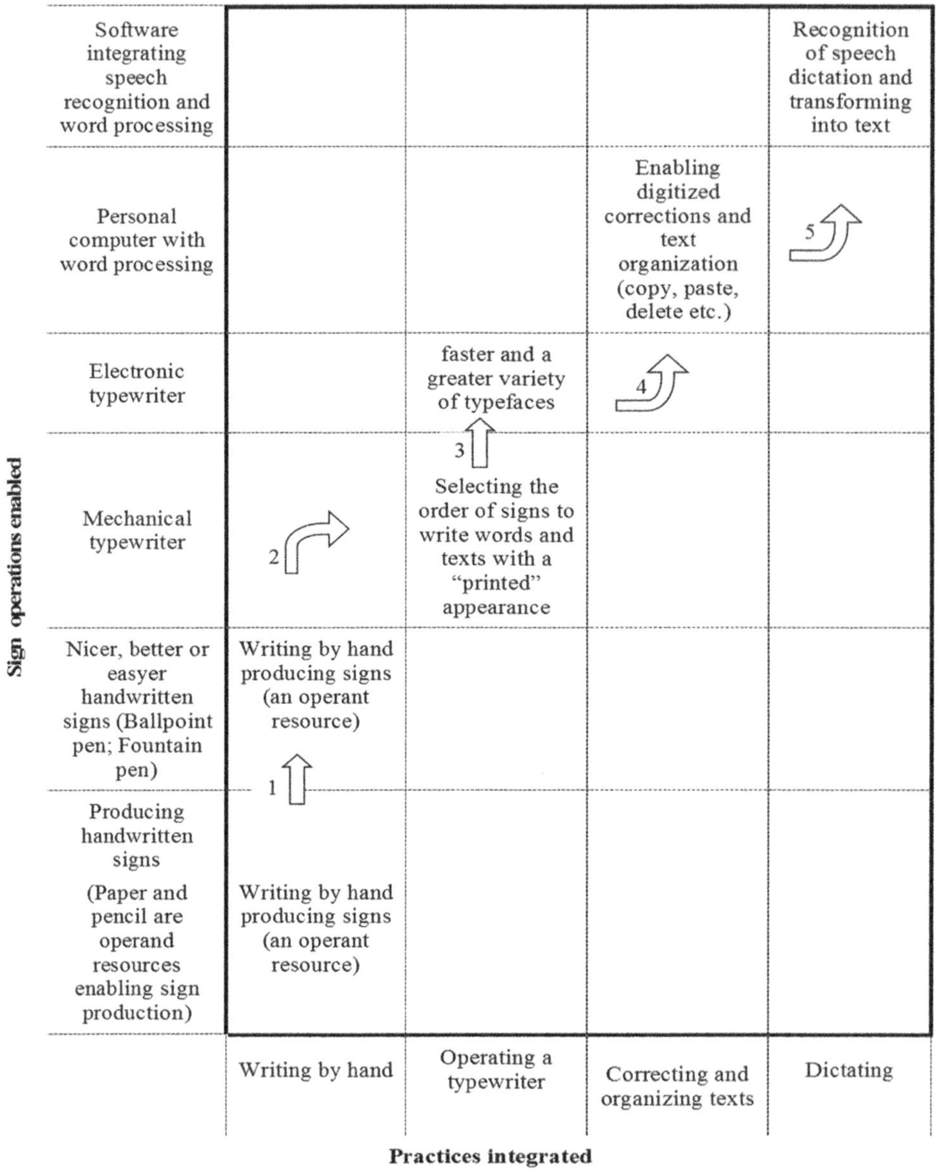

Sign operations enabled		Writing by hand	Operating a typewriter	Correcting and organizing texts	Dictating
Software integrating speech recognition and word processing					Recognition of speech dictation and transforming into text
Personal computer with word processing				Enabling digitized corrections and text organization (copy, paste, delete etc.)	5
Electronic typewriter			faster and a greater variety of typefaces	4	
Mechanical typewriter		2	3 Selecting the order of signs to write words and texts with a "printed" appearance		
Nicer, better or easyer handwritten signs (Ballpoint pen; Fountain pen)		Writing by hand producing signs (an operant resource)			
Producing handwritten signs (Paper and pencil are operand resources enabling sign production)	1	Writing by hand producing signs (an operant resource)			

Practices integrated

Figure 3.7 Word processing innovation pathway. Signs and Practices as Resources in IT-Related Service Innovation from Löbler & Lusch, 2014.

personal computers with (first primitive and later advanced functions, as continuous innovations) word-processing software (indicated by arrow 4 in Figure 3.7). Although the practice of typing remained, the whole practice of writing changed as it enabled copy, paste, and delete, etc., behaviors. Texts now could be rearranged easily, and the correction of a misspelling was no longer issue. Next, a change arrived with speech recognition and the evolution is translating software, which is still primitive, and many design challenges need to be overcome because of implicit knowledge, and with that implicit language, practices. Löbler and Lusch (2014) also discuss the case of video games.

Service innovations must positively affect people's everyday activities and actions, which are based on implicit practices that provide meaning. Unsuccessful digital service innovations are those that do not sufficiently enhance practices and thus are not viewed as valuable. Signs themselves are not the pathway to service innovation in the digital era. However, developing an understanding, through research or other means, of the explicit coordination by using signs through abstract sign systems connected by enhancing the implicit practices of human actors can provide many ideas that can form the basis of digital service innovation. This is especially the case when digital artifacts are embedded in a larger ecosystem that is constantly shifting (Kallinikos et al. 2013). This can be further supported by viewing a practice, in this ecosystem, as a complex adaptive system (Burford 2011).

According to the service-dominant logic perspective, people should not begin with the things they are trying to invent or innovate around, because this conveys the goods-dominant perspective (Paswan et al. 2009), which is neither user centric (Michel et al. 2008) nor practice centric. Service-dominant logic begins with a focus on the actor as a social and economic resource-integrating actor who always cocreates value with other actors but who also uniquely and phenomenologically determines value from their perspective (Vargo and Lusch 2004, 2008). Because, as we have argued, practices are resources, it follows that actors integrate practices. No practice stands on its own but lies in a web of social and economic networks interfacing and integrating to varying extents with other practices.

Consequently, this process begins to create a canvas for service innovation and how it emerges and proliferates aligns with the call by Kallinikos et al. (2013) to view digital artifacts as editable, reprogrammable, interactive, and distributable.

3.4 CONCLUSIONS AND OPEN QUESTIONS

It has been shown that signs render different kinds of service when understood as a complement to practices. Either the practice of language use (the first section) or the practice of coordinating activities (the second section) or coordination thought leading to inventions (the third section) (Löbler 2010; Löbler and Lusch 2014). Signs have emerged as separate entities of inquiry. However we tried to show that signs are deeply embedded in practices. Firstly in practices of language use secondly practices of activities outside of language and thirdly in practices of thoughts.

In language use signs and words do not carry meaning in themselves although they are experienced as they would. I tried to reveal that meaning can be understood as emerging from recursive operations as eigenforms, stable states in recursive operations. Thereby meaning is not fully explicable. There is always a hidden, implicit or tacit part of meaning. Hence the meaning which seems to be connected to a single word can be very different depending on the recursive operations going on. One cannot claim "the right" meaning of a word. It is always a result of social recursive operations which we experience as discourses or alike.

When signs coordinate activities different social groups or communities use them differently. Experts have their jargon which is often very specific and mostly neither intelligible for experts of other fields nor for lay actors. To assume that signs and words carry clear explicit meaning can have harmful potential and can result for example in aircraft crashes. The more the societies are developed and functionally differentiated the more options for "misunderstanding" are there. Whether there is an optimal degree of functional differentiation in a society is still an open question. Has the division of labor gone too far? Do societies need more time and other resources to coordinate and bring together than division of labor has saved? Are humans in the modern digital world reduced to two senses, the audio and visual sense? Which are the consequences for education and the development of the other senses?

In line with Wittgenstein we argued that thinking also depends on signs. In this sense thinking is an operation using signs. Here it is also useful to understand the connections of thinking a practice in a twofold way. Firstly thinking itself is a practice secondly thinking is connected to practices of language use and practices of activities out of language. Often we are not aware that our thoughts are practices and thereby get a more routinized character. One may think of all the routinized thoughts and activities on a normal day. To reflect on these routines and the practices they are performing is a huge source for innovation and can improve our lives.

REFERENCES

Araujo, L.; Kjellberg, H.; Spencer, R. (2008): Market practices and forms: Introduction to the special issue. *Marketing Theory* 8(1): 5–14.

Barker, C. (2012): 10 Deadliest Air Disasters Caused By Miscommunication By. Available online at http://alizul2.blogspot.de/2012/10/10-deadliest-air-disasters-caused-by.html.

Baskinger, M. (n. d.): Visual Noise in Product Design: Problems + Solutions. Available online at http://com119.tripod.com/Visual_Noise.pdf.

Bateson, G. (1970): Form, substance and difference. *General Semantics Bull.* 37: 5–13.

Baudrillard, J. (1975): The Mirror of Production, translated by Poster M (Telos Press, St. Louis): [Orig. pub. 1973.].

Berry, L.; Shankar, V.; Parish, J.; Cadwallader S.; Dotzel T. (2006): Creating new markets through service innovation. *MIT Sloan Management Rev.* 47(2): 56–63.

Burford, S. (2011): Complexity and the practice of Web information architecture. *J. Amer. Soc. Inform. Sci. Tech.* 62(10): 2024–2037.

Cainelli, G.; Evangelista, R.; Savona, M. (2006): Innovation and economic performance in services: A firm-level analysis. *Cambridge J. Econom.* 30(3): 435–458.

Cherrier, H.; Murray, J. (2004): The sociology of consumption: The hidden facet of marketing. *J. Marketing Management* 20(5/6): 509–525.

Cushing, S. (1995): Pilot–Air Traffic Control Communications: It's Not (Only) What You Say, It's How You Say It. *Flight Safety Digest* 14(7), pp. 1–10.

Derrida, J. (1976): Of grammatology. (trans: Spivak, G. C.): Baltimore: John Hopkins University Press (Original work published 1967).

Derrida, J. (1977): Signature event context. *Glyph*, 1, 172–197.

Derrida, J. (1978): Writing and difference. (trans: Bass, A.): Chicago: The University of Chicago Press.

Dorner, N.; Gassmann, O.; Gebauer, H. (2011): Service innovation: Why is it so difficult to accomplish? *J. Bus. Strategy* 32(3): 37–46.

Edvardsson, B.; Olsson, J. (1996): Key concepts for new service development. *Service Indust. J.* 16(2): 140–164.

Engberg, M. (2010): Aesthetics of visual noise in digital literary arts. *Cybertext Yearbook.* Available online at http://cybertext.hum.jyu.fi/.

Erickson, T. (1990): Working with interface metaphors. Laurel B, ed. *The Art of Human-Computer Interface Design* (Addison-Wesley, Reading, MA): 65–73.

Espinosa, J. A.; Lerch, F. J.; Kraut, R. E. (2004): Explicit versus implicit coordination mechanisms and task dependencies: One size does not fit all. E. Sales S. M. Fiore (Eds.): *Team cognition: Understanding the factors that drive process and performance* (pp. 107–129): Washington, DC: American Psychological Association.

Fichman R.; Dos Santos, B.; Zheng, Z. (2014): Digital innovation as a fundamental and powerful concept in the information systems curriculum. *MIS Quart.* 38(2): 329–353.

Fitzsimmons, J.; Fitzsimmons, M. (2000): New Service Development: Creating Memorable Experiences. Sage: Thousand Oaks, CA.

Füllsack, M. (2012): Information, meaning and eigenforms: in the light of sociology, agent-based modeling and AI. Information, 2012(3): 331–343.

Gadamer, H. (1990): Hermeneutik I. Wahrheit und Methode. Grundzüge einer philosophischen Hermeneutik. 6. erw. Auflage: Tübingen: Mohr (Siebeck).

Gallouj, F. (2002): Innovation in the Service Economy: The New Wealth of Nations (Edward Elgar, Cheltenham, UK).

Gallouj, F.; Suvana, M. (2009): Innovation in services: A review of the debate and a research agenda. *J. Evolution. Econom.* 19(2): 149–172.

Giddens, A. (1984): The constitution of society: Outline of the theory of structuration. Cambridge: Polity Press.

Giddens, A. (1991a): Modernity and Self-Identity: Self and Society in Late Modern Age: Polity Press, Cambridge, UK.

Giddens, A. (1991b): The Consequences of Modernity. Cambridge: Polity Press.

Grönroos, C. (2008): Service logic revisited: Who creates value? And who co-creates? *Eur. Bus. Rev.* 20(4): 298–314.

Habermas, J. (1985a): The theory of communicative action: Vol. 1. Reason and the rationalization of society. Boston: Beacon Press.

Habermas, J. (1985b): The theory of communicative action: Vol. 2. Lifeword and system: A critique of functionalist reason. Boston: Beacon Press.

Kallinikos, J.; Aaltonen, A.; Marton, A. (2013): The ambivalent ontology of digital artifacts. *MIS Quart.* 37(2): 357–370.

Kaufmann, L.H. (2003): Eigenforms – objects as tokens of eigenbehaviour. *Cybernetics and Human Knowing*, 10(3–4): 73–89.

Kaufmann, L.H. (2005): Formal Systems Eigenform. *Journal Kybernetes*, 34(1–2): 129–150.

Löbler, H. (2010): Signs and Practices: Coordinating Service and Ralationships. *Journal of Business Market Managment*, pp. 217–230.

Löbler, H. (2011): Position and potential of service-dominant logic—Evaluated in an "ism" frame for further development. *Marketing Theory* 11(1): 51–73.

Löbler, H. (2014): When Trust Makes It Worse—Rating Agencies as Disembedded Service Systems in the U.S. Financial Crisis. In *Service Science* 6(2), pp. 94–105.

Löbler, H.; Lusch, R.F. (2014): Signs and Practices as Resources in IT-Related Service Innovation. In *Service Science* 6(3), pp. 190–205.

Löbler, H.; Wloka, M. (2015): Loyalty' Between Talk and Action – Meaning as Eigenforms of Recursive Operations. Bartsch, S.; Blümelhuber, C. (Eds.): Always ahead: Ideen für das Marketing: Springer Gabler.

Luhmann, N. (1995): Social Systems. Stanford Calif: Stanford Univ. Press (Writing science).

Luhmann, N. (1996): Membership and Motives in Social Systems. *Systems Research* 13(3), pp. 341–348. Available online at http://search.ebscohost.com/login.aspx?direct=true&db=bth&AN=18506048&site=ehost-live.

Merriam Webster (2014): Online Lexicon. Retrieved December 5, 2014, from http://www.merriam-webster.com/dictionary.

Michel, S.; Brown, S.; Gallan, A. (2008): An expanded and strategic view of discontinuous innovations: Deploying a service-dominant logic. *J. Acad. Marketing Sci.* 36(1): 54–66.

Nambisan, S.; Sawhney, M. (2007): The Global Brain: Your Roadmap for Innovating Faster and Smarter in a Networked World. Prentice Hall: Upper Saddle River, NJ.

National Transportation' Safety Board Washington, D.C. 20594 (1990): Aircraft accident report avianca, the airline of Columbia Flight: Avianca 052. Washington D.C.

Nijssen, E.; Hillebrand, B.; Vermeulen, P.; Kemp, R. (2006): Exploring product and service innovation similarities and differences. *Internat. J. Res. Marketing* 23(3): 241–251.

Ogden, C.; Richards, I. (1946): The Meaning of Meaning. A study of the influence of language upon thought and of the science of symbolism. 8. Ed. New York: Harcourt, Brace & World, Inc.

Ordanini, A.; Parasuraman, A. (2011): Service innovation viewed through a service-dominant logic lens: A conceptual framework and empirical analysis. *J. Service Res.* 14(1): 3–23.

Paswan, A.; D'Souza, D.: Zolfagharian, M.A. (2009): Toward a contextually anchored service innovation typology. Decision Sci. 40(3):513–540.

Reckwitz, A. (2002): Toward a theory of social practices: A development in culturalist theorizing. *Eur. J. Soc. Theory* 5(2): 243–263.

Sawhney, M.; Verona, G.; Prandelli, E. (2005): Collaborating to create: The Internet as a platform for customer engagement in product innovation. *J. Interactive Marketing* 19(4): 4–17.

Shannon, C. (1998): The mathematical theory of communication. Shannon, C.; Weaver, W., eds. The Mathematical Theory of Communication (University of Illinois Press, Urbana): 29–125. [Orig. published 1949].

Schatzki, T. (1996): Social practices: A Wittgensteinian approach to human activity and the social. Cambridge: Cambridge University Bridge.

Schatzki, T. (2007): Introduction. Human Affairs 17(2): 97–100.

Schau, H.; Muñiz, A. Jr; Arnould, E. (2009): How brand community practices create value. *J. Marketing* 73(5): 30–51.

Smith, D.; Kirby, C.; Kimball, R.; Harslem, E. (1982): The star user interface: An overview. *Proc. AFIPS National Comput. Conf.* ACM: New York: 515–528.

Sheehan, J. (2006): Understanding service sector innovation. *Comm. ACM* 49(7): 43–47.

Sheth, J.; Sisodia, R.; Sharma, A. (2000): The antecedents and consequences of customer-centric marketing. *J. Acad. Marketing Sci.* 28(1): 55–66.

Spencer-Brown, G. (1969): Laws of form – Gesetze der Form (4th ed.). Leipzig: Bohmeier Verlag.

Swanson, E. (1994): Information systems innovation among organizations. *Management Sci.* 40(9): 1069–1092.

Toups, Z. O.; Kerne, A. (2007): Implicit coordination in firefighting practice: Design implications for teaching fire emergency responders. *Proceedings of the SIGCHI conference on Human factors in computing systems* (pp. 707–716). CHI: San Jose, Californien.

von Foerster, H. (1981): Objects: tokens for (eigen-) behaviors. In Heinz von Foerster (Ed.): Observing systems. Seaside, Calif.: Intersystems Publications (Systems inquiry series), pp. 274–285.

von Foerster, H. (2003): Understanding understanding: Essays on Cybernetics and Cognition. New York: Springer.

von Krogh, G.; Haefliger, S.; Spaeth, S.; Wallin, M. (2012): Carrots and rainbows: Motivation and social practice in open source software development. *MIS Quart.* 36(2): 649–676.

Vargo, S.; Lusch, R. (2004): Evolving to a new dominant logic for marketing. *J. Marketing* 68(1): 1–17.

Vargo, S.; Lusch, R. (2008): Service-dominant logic: Continuing the evolution. *J. Acad. Marketing Sci.* 36(1): 1–10.

Vargo, S.; Lusch, R. (2011): It's all B2B ... and beyond: Toward a systems perspective of the market. *Indust. Marketing Management* 40(2): 181–187.

Venkatesh, A.; Penaloza, L.; Firat, A. (2006): The market as a sign system and the logic of the market. Lusch, R.; Vargo, S., eds. *The Service-Dominant Logic of Marketing, Dialog, Debate and Directions*. M.E. Sharpe: New York: 251–265.

Warde, A. (2005): Consumption and theories of practice. *Journal of Consumer Culture* 5(2): 131–153.

Weaver, W. (1998): Some recent contributions to the mathematical theory of communication. Shannon, C.; Weaver, W., eds. *The Mathematical Theory of Communication*. University of Illinois Press: Urbana: 1–28 [Orig. published 1949].

Wittgenstein, L. (1960): The Blue and Brown Books, 2nd ed. Harper Row: New York.

Wittgenstein, L. (2008): Philosophical Investigations: The German Text, with a Revised English Translation. Blackwell: Oxford, UK.

Yoo, Y.; Henfridsson, O.; Lyytinen, K. (2010): The new organizing logic of digital innovation: An agenda for information systems research. *Inform. Systems Res.* 21(4): 724–735.

Chapter 4

Challenges to management accounting in the new paradigm of service

Lino Cinquini & Andrea Tenucci
Institute of Management, Scuola Superiore Sant'Anna, Pisa, Italy

SUMMARY

The last decades show an increasing percentage of the service industry (or tertiary sector) on GDP in most countries. Services are also infusing into products, giving rise to product servitization. In the meantime on the academic side, Service-Dominant logic and Service Science are emerging in marketing and management areas. Despite this, research in management accounting has maintained its focus largely on metrics and approaches manufacturing-oriented.

The Chapter aims at providing insights and addressing the main challenges for management accounting by considering its role with respect to contemporary changes in the service paradigm. In particular, the Chapter depicts six critical aspects or dimensions of business service-related change in which the approach of the traditional Goods-Dominant Logic perspective differs from that of Service-Dominant Logic, and the implications for management accounting objects and drivers to be emphasized and focussed by an effective measurement.

The evidence and reflections of the Chapter suggest the existence of many under-explored and relevant issues and critical areas that challenge service-oriented research in management accounting, such as: the measurements of value in a relationship-based value creation, new drivers of profitability, new customer centric metrics, capacity-related cost drivers and the role of costing, capacity management and business modelling for pricing. The Chapter closes by suggesting the innovations and trends of utmost importance to be considered in management accounting to design and implement measurements and tools in service contexts.

Keywords

Management accounting, service management, service-related change, Service-Dominant Logic

4.1 INTRODUCTION

The diffusion of new technologies and the new opportunities emerging from their application have boosted the relevance of "service" far beyond the increase of the weight of "service sectors" in the overall economy. A different perspective around the essence of contemporary business has emerged: "Service-Dominant Logic" (SDL) (Vargo and Lusch, 2004; 2008) has been proposed in marketing studies, and the process of "servitization" is expanding as a competitive strategy in manufacturing (Vandermerwe and Rada, 1988; Oliva and Kallenberg, 2003). More recently, "Service Science" has been proposed as a set of intertwined disciplines able to face the complexity of social and economic organizations (service systems) (Maglio and Spohrer, 2008; Spohrer and Kwann, 2009). Further, service orientation has enhanced the importance of firm interrelationship and supply chain for competitive strategy. Service Supply Management has become relevant not only for the increasing importance of service industry in the world economy, but also for manufacturing companies which achieve more revenue from services linked to physical goods (Lin et al., 2010).

Management Accounting (MA) is recognized as the discipline dealing with the provision of relevant accounting information for business decision making and organizational control (Horngren et al., 2001; Garrison et al., 2008; Drury, 2012) and despite these major changes, its research and principles largely refer to the paradigm of manufacturing. Notwithstanding progress in Services, they are generally considered as "special products" in costing, performance, decision-making and control, and service organizations has been underestimated as relevance and generally treated as special case of manufacturing industries. Although in the last decade research in MA has been carried out on the role of accounting in value chain analysis and inter-organizational relationships (Caglio and Ditillo, 2008) and on the diffusion of innovative management accounting tools (Langfield-Smith, 2008; Cinquini and Tenucci, 2010), the implications for MA of recent developments in topics such as servitization, SDL and Service Science have not been explored in deep yet.

This Chapter aims at providing insights into the main challenges for management accounting and its role with respect to contemporary change in service as emerging by recent research. It then suggests research agenda and identifies six topics that challenge MA in providing valuable information for business according with the principles argued by SDL and Service Science.

The relevance of moving research in this direction has been recently recognized in service literature. Among the ten overarching research priorities for the Science of Service, Ostrom et al. (2010) indicate the *"Measuring and optimizing the value of service"*. The authors recall the need of research in this specific field indicating six topic areas where further research is particularly needed: they particularly address the lack of cost accounting systems of service providers in tracking service costs across all functions and business units and the need for a cross-disciplinary approach (where marketing and operations but also accounting, finance and information technology are included) in order to catch the value creation as it crosses functional boundaries. More broadly, some authors have started a reflection on the relation between Accounting and Service Science (Kerr, 2008).

The chapter is structured as follows: Section 2 introduces the evolution of the service framework and the related emerging issues for MA. Section 3 deepens the potential roles of MA under SDL and Service Science. Then Section 4 summarizes the role of MA for service and looks ahead with some considerations in this stream. The Chapter then closes with Section 5 with some final reflections on the role MA can play under a Service-dominant mindset.

4.2 THE PILLARS OF SERVICE EVOLUTION AND THE IMPLICATIONS FOR MANAGEMENT ACCOUNTING

Research on services is historically based on features distinguishing between services and physical goods. Generally, a distinction that constitutes a key reference point is ascribable to Shostack (1977), who described the differences between products (goods) and services from the marketing management perspective in four aspects: intangibility, heterogeneity, inseparability and perishability—from which the acronym IHIP derives.

According to contemporary management and marketing literature, three emerging pillars are building a new service framework in modern business: the servitization process, the Service-Dominant Logic and the discipline of Service Science.

4.2.1 Servitization

The first pillar, the servitization process, has been highlighted by Vandermerwe and Rada (1988) referring to the trend related to "*the increased offering of fuller market packages or 'bundles' of customer focussed combinations of goods, services, support, selfservice and knowledge in order to add value to core corporate offerings*" (Vandermerwe and Rada, 1988). They argue that there are at least three reasons why manufacturing firms should servitize – (i) to lock out competitors by avoiding price competition and raising barriers; (ii) to lock in customers raising the costs of substitution and (iii) to increase the level of differentiation. A recent contribution by Neely (2008) defines "servitization" as involving "(...) *the innovation of an organisation's capabilities and processes so that it can better create mutual value through a shift from selling product to selling Product-Service Systems*". Increasing research has been carried out about the issues faced by manufacturing companies in servitizing their production and the strategic and managerial implications of this process (Mathieu, 2001; Oliva and Kallenberg, 2003; Brax, 2005; Gebauer and Friedli, 2005). Research about the economic impact of servitization has shown a "service paradox" related to the difficulty in gaining the expected level of returns from services (Gebauer and Friedli, 2005). While servitized firms generate higher revenues, they tend to generate lower net profits as a % of revenues than pure manufacturing firms. Although these researches present a limitation in that firm performance could be differently constructed (i.e. profitability could be measured comparing profit to what is invested in the firm, either total assets or stockholder equity), recent findings based on empirical research have addressed the reason that servitized firms actually have higher average labour costs, working capital and net assets (Neely, 2008; Neely et al., 2011).

Many authors however maintain that the servitization process offers advantages for both suppliers and customers. From a supplier perspective, servitization is a way to increase sales revenues, while from a customer perspective servitization reduces risks and decreases uncertainty making predictable maintenance and support costs (Slack, 2005).

4.2.2 Service-dominant logic

The second pillar of the emerging service framework has occurred in marketing literature since 2004. Vargo and Lusch (2004; 2008) proposed a significant change, erupting in a "new dominant logic" for the theory and practice of marketing: Service-Dominant Logic. In their seminal article, Vargo and Lusch addressed the service (rather than the product) as what creates value for the customer; accordingly, goods are interpreted as mere means or delivery mechanisms of service provision. SDL says that the application of competences for the benefit of another party—that is, service—is the foundation of all economic exchange: even when goods are involved, what is driving economic activity is service-applied knowledge. Thus, service is the basis of all social and economic exchange; all businesses are service businesses and all economies are service economies (Vargo and Lusch, 2008). The value for customers emerges within the customers' sphere for every kind of consumption: it is the *value-in-use* in their value-generating processes. This perspective challenges the prevailing view that value for customers in goods is embedded in the outputs of firms' manufacturing processes and expressed as *value-in-exchange* (Goods-Dominant Logic, GDL). Service, therefore, becomes the general case and the common denominator of the exchange process: service is always traded, while the goods, when used, are the supports for the process of service delivery (Normann and Ramirez, 1993; Normann, 2001; Vargo and Lusch, 2008; Grönroos, 2006). The interpretation of the *value creation process* changes accordingly: value is not created by the provider but by the customer in its value-generating processes, i.e. value is created when customers *use* goods and services (*value-in-use*) rather than being embedded in goods or services (*value-in-exchange*). This way, customers become value co-creator because the value is generated by the consumption of an offer (good or service); the offer constitutes the provision by firms of the necessary resources for the value-generating processes by customers. On the other hand, in doing so firms have the possibility to "enter" the consumption processes and to develop opportunities to co-create value with their customers (Grönroos, 2008; Vargo and Akaka, 2009; 2012).

4.2.3 Service science

The third pillar of the emerging service framework is constituted by the Service Science discipline. IBM began a major reflection on structural change in business scenario of the new millennium: pivoting on service as a central subject in modern business settings, IBM proposes an integration of disciplines, scientific, managerial and engineering (Service Science) aimed at offering effective solutions to contemporary issues of enterprise. The basic objects of the new discipline in the new context become the systems of "services" (service systems) (IBM, 2004; Spohrer and Kwann, 2009). In line with the Service-dominant worldview the service system is viewed as the fundamental abstraction of the study of value co-creation in service science (IBM and IfM, 2008). *"Service systems are physical symbol systems that compute the changing value of knowledge in the global service system ecology (...). Basically, all these entities can be viewed as service systems that depend on value co-creation interactions to survive generation after generation and improve the quality of life of the people inside them by inventing better value co-creation mechanisms for larger or smaller populations of entities in diverse environments. Typically they involve people, technology, organisations and shared information (also summarised as individuals, infrastructure, institutions and information)."* (Spohrer et al., 2013: p. 9).

4.2.4 Implications for management accounting

From the management accounting perspective, services have been traditionally considered as "special products" (with IHIP features) and studied according to their differences respect to manufactured goods. The advent of innovative perspectives asks for a new approach to management accounting in designing appropriate managerial information. However, the recalled shifts in service concepts and insights into the core of value creating process involve important changes and new relevant questions for business decision-making and information provided by Management Accounting Systems, such as:

* what are the consequences of servitization on the relevant accounting information to support decision making of a product manufacturer becoming a service provider?
* how to measure the value created in the process of value co-creation and performance considering integration of resources with networked partners?
* how to consider properly value-in-use to determine value-in-exchange and thus also for pricing?
* is there any shift in "output as a relevant accounting object" towards the increasing importance of the "accountability of the consumption process" by customer?
* what are the consequences, in terms of management accounting and control, of the shift to a process-driven and service-centric logic that provides a more solid foundation for a transition to a service-provider model (Vargo and Lusch, 2008)?

As we will acknowledge in the Chapter, advancements in cost and performance measurement techniques (such as Activity-Based Costing, Strategic Management Accounting, Balanced Scorecard) have constituted and still represent a potential answer to the service shifts in order to innovate and improve management accounting systems. On one side, these techniques are based on the analysis of activities and processes that are key elements for the government of services. On the other side, non-financial metrics integrate financial measures and allow to identify more effectively performance drivers and to evaluate results (Brimson and Antos, 1994; Kaplan and Norton, 2008).

Nevertheless, a focus on Management Accounting position with respect to these recent developments in marketing and management research has not been clearly appointed yet, nor a reflection on the consequences on MA in considering the pervasive process of the infusion of service in modern business, the impact of digitization in firms structures and new management and information needs. The impact of user-interaction in collecting and managing information only recently has started and new directions for an innovative framework in this respect have been proposed (Bhimani, 2006; Bhimani and Bromwhich, 2010; Laine, 2009; Cugini et al., 2007; Coller et al., 2011; Hilker, 2013).

4.3 THE ROLE OF MANAGEMENT ACCOUNTING IN THE SERVICE ENVIRONMENT: IN SEARCH OF NEW DIRECTIONS

In accounting literature Management accounting systems are recalled to satisfy a crucial role in the company, by providing (financial and non-financial) information to assist managers in their activities. In particular management accounting supports three managerial activities: planning, controlling and decision making (Garrison et al., 2008). *Planning* involves establishing goals and specifying how to achieve them. *Controlling* involves gathering feedback to ensure that the plan is being properly executed or modified as circumstances change. *Decision making* involves selecting a course of action from alternatives. Management accounting is concerned with collecting, classifying, processing, analysing and reporting information to managers. Unlike the financial accounting information prepared for external purposes and addressed to external stakeholders, management accounting information is designed to help internal decision makers (managers) for different purposes.

In this section of the Chapter an investigation is proposed, starting from the management accounting framework, to explore the challenging issues that rise in it as a consequence of changes in service economy and the emergence of SDL and Service Science business perspectives.

4.3.1   Cost accounting issues considering the emerging service framework

Cost information is traditionally one of the most relevant information provided by MA. There are many research questions arising on the cost-side when investigating service companies. Which are the potential cost objects of the analysis? And is the output (product or service) still a meaningful cost object? Which costing tool, if any, becomes more relevant? Which are the drivers of the price-cost relationship? How to measure the value in-use and value in-exchange? How does profitability relates to value for customer?

4.3.1.1   Costing and value

To distinguish between direct and indirect costs, we previously need to define a cost object as anything to which cost information is desired (Horngren et al., 2001; Garrison et al., 2008; Drury, 2012). The distinction between direct and indirect costs highlights the issue of determining which potential cost objects in service companies are worthwhile to be considered. The cost object may be the service (mailing, consulting, e-mail, web hosting, etc.), the customer or the end user, if different. Anyway the issue of cost allocation increases in service companies as many cost objects can be identified. The R&D, IT and infrastructure costs, among the others, are not only fixed but also indirect costs. In this respect, we can conclude that in service companies we can reasonably expect to find greater part of the costs as indirect (Dearden, 1978; Modell, 1996).

SDL addresses new relevant objects that open new perspective for costing. In this vein, it is worthy of consideration a recent paper by Ng et al. (2012), that presents a visualisation of Rolls-Royce offering from a SDL perspective in equipment-based service. The case of Rolls-Royce is presented as an avenue through which to explore an alternative view of the firm's value proposition, a visualisation informed by SDL that could aid organisations in their transition from GDL to SDL. The study finds that the SDL visualisation of the firm's value proposition is possible by the identification of *value-creating activities* (VCAs) towards value-in-use equipment-based service.

Understanding these VCAs/attributes enables the firm to construct its value proposition around the value-in-use realized by the customer. As a consequence, costing the VCA is particularly important, because it makes possible to estimate the cost of different bundles of VCAs through the simulation models of the processes supporting each value-creating activity. This way, the most efficient bundle to provide from the perspective of the firm's resources and costs can be determined. Ultimately, this Chapter suggests the VCAs provided for the customer, informed by the SDL view of the firm's value proposition, as new cost objects. This cost information supports the selection of the VCA bundle that most effectively contributes to the customer system and the efficiency of firm's delivery. This perspective in costing may be considered an extension of the Target Costing approach in determining the "value index" in costing product "function" (Yoshikawa et al., 1993).

In this context MA can find certain coherence and a wider application in Performance Based Contracting (PBC) strategies. PBC is known variously as "Performance-based logistics" (PBL) in defense contracting, "Outcome Based Contracting" (OBC), "Performance-based life-cycle product support", "Contracting for Availability" in aftermarket sales or "Power-By-The-Hour" in the commercial sector. It has emerged as a strategy for improving the performance and lowering the costs of complex network in the post production phase. In other words there is a supplier network, responsible for the performance of the system, compensated on the base of its ability to deliver a performance-based outcome instead of being paid to overhaul parts or provide replacement components (Randall et al., 2010). Not surprisingly both service, in SDL, and performance, in PBC, use a combination of goods, services, skills, and knowledge to achieve a performance outcome. Performance and service represent the value proposition offered by the network or service ecosystem (Randall et al., 2010).

PBC is an approach representing a supply chain application of SDL where MA can contribute with providing cost information crucial for the network. Under this perspective the cost of installing the component under the view of the single company is losing relevance, whereas it becomes more important to have a wider perspective of costs and revenues of the entire network or service ecosystem. Here MA has to be extended moving toward the Strategic Cost Management (SCM) perspective (Shank and Govindarajan, 1993) where the entire supply chain is considered as object of analysis. SCM represents the joint application of financial analysis with value chain analysis, strategic positioning analysis and cost driver analysis. Strictly linked with SCM some authors, in the accounting literature, also recall the Interorganizational Cost Management (IOCM) as "*a structured approach to coordinating the activities of firms in a supplier network so that the total costs in the network are reduced*" (Cooper and Slagmulder, 1999). In this context both SCM and IOCM could represent adequate tools to monitor costs and revenues within the network or service ecosystem and serve as facilitator to reach the goal of the PBC.

If considering the perspective of value co-creation and servitization, it is no longer relevant the "cost of production/product" or "service", but the "cost of use" becomes relevant as part of its overall life cycle. In other words, the focus shifts from the cost of provider to the analysis of the costs of the service offered and its maintenance over its life cycle. The before mentioned diffusion of Outcome-based contracting, a contracting mechanism that allows the customer to pay only when the firm has delivered outcomes rather than merely activities and tasks, is somehow related with the relevance of information about the "cost of use" by customer and the importance of this issue in MA. An example is the case for Rolls Royce "Power-by-the-hour" contracting for the service and support of their aerospace engines, where the continuous maintenance and servicing of the engine is not paid according to the spares, repairs or activities rendered to the customer, but by how many hours the customer gets power from the engine (Ng et al., 2010). OBCs have been shown to provide cost efficiencies as both the firm and the customer's objectives become much more aligned.

In this perspective, the costing systems capable of detecting the "Total Cost of Ownership" (TCO) (or the Life cycle cost – LCC) are becoming increasingly important. The shift in the object of cost analysis (from the product to the user) can support decisions aimed at reducing the utilization costs for the customer through the innovation in the design of the offer. Thus, the cost/performance ratio may increase and therefore the value for the customer, which can also be monetized with a reduction in cash outflow to be paid to the supplier.

The TCO analysis was born with reference to a more accurate assessment of the cost of supply within the supply chain that enables to understand the burden beyond the transaction price (Ellram, 1993; Ellram and Siferd, 1998). Such an approach, however, can also be applied with respect to the final customer, to understand the nature and effectiveness of services provided by the manufacturer/supplier and act in terms of both performance improvement on that of efficiency in the perspective of value co-creation. An example in this sense is provided by Barontini et al. (2013) with the case of ElsagDatamat, a company of the Finmeccanica Group specialized in the design and development of systems, products, solutions and services for hi-tech automation, logistics and mobility and ICT. The company uses TCO in customer relationship in order to demonstrate the economic benefit resulting from assigning some services to them.

4.3.1.2   Costing in service offerings

Furthermore, recent research on servitization and its impact on performance of manufacturing firms (Neely, 2008; Neely et al., 2011; Visnjic et al., 2012) reveals some of the most important drivers to be monitored and managed in renovated management accounting systems. The monitoring of the complexity of service offerings in their cost side becomes fundamental, and more sophisticated costing systems are needed to ascertain the efficiency of their provision. Visnjinc et al. (2012), through an extensive empirical research of financial dataset, show that servitization can be seen as a good growth strategy, but with limitations to what can be seen as profitable service growth. The results suggest that extending the *breadth* of service offering by expanding in the spectrum of service portfolio may result in diminishing efficiency, i.e., that firms that intend to grow by expanding the scope of service offering, adding more and more services to the portfolio, may realize inferior profit margins. The development of Time-Driven Activity-Based Costing System (Kaplan and Anderson, 2007) seems suitable in catching cost information on service, given its ability to reflect the pattern of resource consumption using the main human-related cost driver, the time, and being relatively easy to design and implement. Regarding servitization, it is worthy of consideration a recent paper by Laine et al. (2012) claiming for a "role of MA in supporting servitization". Three major roles of MA are identified to this aim: *justifying* (Why do we desire servitization?), *defining* (What does servitization mean to us?) and *controlling* (How does the servitization proceed/affect us?). These processes may concern three relevant units of analysis, implying potentially different accounting objects for servitization: service logics at the corporate level according with the "service-dominant logic" (Vargo and Lusch, 2004); service as a solution for the customer at the customer and product levels (Grönroos, 2000 and 2008); the IHIP characteristics as differentiators between goods and services (Shostack, 1977). An analysis on the potential role of MA is then developed, considering the different MA task applied on the different viewpoints and using the uncertainty framework used of Burchell et al. (1980), Hopwood (1980) and Chapman (1997) to explain the roles of MA in different decision-making situations.

4.3.1.3   Costs and revenues linkage

A further important issue concerns the dissociation between investments (costs) and sources of revenue, which raises new problems in the rationale of the traditional approach to "costing for pricing" (Bhimani and Bromwich, 2010). Particularly in contexts where a co-partnership with the customer and open innovation is pursued, what is "produced" is not what generates revenues, so there is no sense in pricing to follow traditional models of the cost-plus or market base type. The pricing is rather linked to the dynamics of business strategy and revenue generation, and it is dissociated from the cost of production. In most cases, the price charged for the service is zero, in some cases the price is fixed and in others the price changes according to customers choices (cannot be set *a priori* – i.e. the cost of Dropbox service depends on the storage space required).

In the cases of Facebook, Google and, in general, web-companies the volume of users becomes the most important driver of revenues and profitability, as it determines the attraction of investment in advertising (i.e. 96% of Google's revenues in 2010 and 2011 and about 94% in 2012 derived from advertising). In this way the conditions for coverage of the amount of investment (fixed costs) in IT infrastructure related to the development of hardware and software applications are created; but the direct link between costs and prices characterizing the business world of the GDL is missing. In the world of goods, the cost-price causal link derives from the direct focus on the processes that create products or services, under the assumption that producers should manage the resources they own.

4.3.2   Performance and value distribution in the "new" emerging service framework

Many other research questions arise to management accounting in service framework. How does the company measure its own performance? Is the performance of the single company anymore important? Is it possible to measure the value co-created by the customer/user?

The financial performance measurement in SDL is recognized critical: "*Although we do indeed argue that value is only created through co-creation and in interaction with the customer, we recognize that monetary flows are critical.*" (Vargo and Lusch,

2006: p. 50). As literature and research argue (Stewart, 1997; Lev, 2001), the transactional nature of financial accounting systems limits the possibility to represent the value created in term of intangibles by investments in marketing or service development.

We believe that in order to have an idea of the value co-created it is necessary for the company to extend the analysis beyond its physical boundaries. Coherently with what we introduced previously, MA can have a role providing some tools such as IOCM or the Value Chain analysis. Both are related to a wider approach–called Open Book Accounting (OBA) defined as *"the systematic disclosure of cost information between legally independent business partners beyond corporate borders"* (Hoffjan and Kruse, 2006, p.40). So the sharing of cost information, traditionally considered extremely "confidential", represents a first step for MA toward the extension of the analysis to the supply chain, network or service ecosystem (Chen, 2011).

The evidence on OBA provided in accounting literature offers a potential bridge between MA and SDL. Carr and Ng (1995) and Mouritsen et al. (2001) provide examples of companies sharing information related to costs (cost structure, cost of materials etc.) and time (set-up time, time to customer or handling time) among the supply chain. Whereas Kajuter and Kulmala (2005) document a wider set of cost information disclosed in a supply network. Under the perspective of SCM, the accounting literature further demonstrates the usefulness of the value chain analysis combined with an activity-based costing approach in supporting interorganizational cost reduction projects (Dekker and Van Goor, 2000; Dekker, 2003). These evidence should be seen as a first effort and step toward a wider evaluation of the co-created value along the supply chain among which the cost information represents only an element. In any case the application and the support of such techniques is a sort of fundamental pre-requisite.

4.3.2.1   Performance in servitization

If considering servitized business environments, the issue of designing a performance measurement systems adapted to capture relevant dimensions of performance for product-service providers is an emerging topic of increasing relevance. Visnjic and Van Looy (2011) have conducted a three-year study in a company in order to understand the nature of service performance in an industrial, product driven global equipment manufacturer. The findings were that three important components are required to measure service business performance effectively: *service coverage,* that addresses the question, "How many – or what proportion – of our (product) customers are also buying services?"; *service realisation,* that addresses the questions "How well are we covering the service needs of those customers?", and the effectiveness of *complementarity* of the product-service relationship that impacts on overall performance. Service coverage indicates what percentage of a firm's (product) customers is also buying services. This ratio measures the extent to which the company has a service contact with its installed base of products. Service realisation is proposed as a ratio between actual service sales and potential sales of the most complete service offering. Beside this, Visnjic et al. (2012) research shows that, due to the nature of the relationship between product and service offering, these two isolated sets of performance measures are insufficient to measure all aspects of performance. Because of the close connection between product and service offerings the overall performance is very likely to be impacted by the effectiveness of the product-service relationship. There may be synergic or conflicting interrelations in product servitizing between product sales, that want to optimise the sales offer by arguing that the product doesn't need servicing, and the service technician and salesman, that can propose endless makeovers of an old machine, even when it is neither beneficial to the customer nor the manufacturer. In this case, an increase in product sales will be associated with a decrease in service sales and vice versa. As a consequence, a metric to measure the relational spillovers between the two activities is required: the authors propose a *complementarity index* by a correlation factor between product and service sales. Finally, an integrated view by a *servitization matrix* to visualize the state of the service business and its relation with the product business by the three metrics is developed. This contribution claims for a different approach to performance required by servitization, as performance measurement systems designed for product-only business models are ineffective. With the increasing of firms adopting a business model that combines a product and service component there is a need for a deep rethinking on performance measurement systems.

4.3.2.2 *Measuring co-created value and pricing*

In the world of services, however, the process of value co-creation is the ultimate source of profitability that should be monitored, measured and managed. In internet-based services (i.e. Social networks or platforms like eBay) the situation seems to be more complicated because of the influence of particular business models. Such services include digital entertainment features but also a personal entertainment linked to an experience that the user lives in the utilization of the service. In the case of Facebook the subscriber, using the platform, can be considered an actor in the co-creation process though which Facebook Inc. is selling advertisements and other marketing services to some enterprises, seen as "real" customers. This is still the *value-in-use* recalled earlier. But in our management accounting perspective the value-in-use implies the issue of measuring the part of the value co-created by the customer/user. For instance, in the case of Facebook, the time spent on the platform, the number of posts, the number of friends of the subscriber are all potential performance indicators explaining and contributing to the co-creation of value.

A major area where the rising issues in service affect the development of management accounting research, concerns the problem of the distribution and measurement of value between value co-producers/co-creators in a system of services. In this respect, the co-creation of value with customers becomes important, both in the business to business relationships and in business to consumer: in the service-oriented logic company's goal is the mutual creation of value for the company itself and for its customers and the service is a mediating factor in this process (Grönroos and Ravald, 2009). In other words, the value that a company can create in the relationship with a customer depends on the value that the same customer can create from the involvement in the relationship. In this sense a "mutual value creation" rises: the customer is acting as co-producer in the process of the supplier while the supplier is acting in the corresponding process of creating customer value and is involved in an active way (Gronroos and Helle, 2010: p. 570). Gronroos and Helle (2010) carried out a step towards the measurement of value in this logic. They propose a model of evaluation in which the joint supplier-customer productivity and how this comes from the efficiency and effectiveness of the relationship itself are considered. It is clear that the capacity to make this measurement depends on the availability of data based on costs and expected cash flows as well as the degree of trust and mutual opening of accounts by the availability of the actors involved in the relationship.

With this in mind and focusing on the value captured by the service provider, Storbacka and Nenonen (2009) suggest that the "value capture" can be measured by discounting the future profits arising from the relationship with the customer, and also argue that this value can be used as a proxy of value creation for shareholders. The value of long-term relationship between customer and supplier (Ravald and Gronroos, 1996) especially in service companies becomes a subject not only for the exclusive use of the marketing field but also an area in which management accounting can make a substantial contribution.

Furthermore, in service systems like web-companies resources are proposed, consumed and value co-created by a web of stakeholders, including customers themselves, that have something to gain and something to give to the system. Here pricing models based on direct exchanges are ineffective, as cause (who is delivering value) and effect (who is consuming value) become blurred. What becomes valuable is the *right of access* such resources which include firm and customer. *Capacity-driven revenue models* are required in order to define the capacity of the system to deliver outcomes and the optimization to match customers' willingness to pay (Ng, 2010). New challenges therefore have to be faced to unbundle the value co-created by web of actors. MA in this respect is required to develop a stronger orientation in managing revenues together with costs.

4.4 MANAGEMENT ACCOUNTING SHIFTING FROM SDL TO GDL: TAKING STOCK AND LOOKING AHEAD

Previous sections have provided innovative aspects of business in the perspective of Service-Dominant Logic and research topics that represent new challenges to MA research. These new trends in business require a major shift in most of the traditional orientations in which management accounting has developed within the branches of its knowledge domain, both in theory and in practice.

In the field of management accounting, as aforementioned, we do recognize the research area of Strategic Management Accounting (SMA) that is more developed toward the use of accounting information for strategic purposes. The strategic aspect of the information provided by SMA stands in the perception of the company outside its physical boundaries and then including information related to customers, competitors, networks or the entire supply chain (Shank and Govindarajan, 1993; Bromwich, 1990).

Research has been carried out in this respect regarding the role of accounting in value chain analysis and inter-organizational relationships (Caglio and Ditillo, 2008), the diffusion of innovative management accounting tools related to SMA (Langfield-Smith, 2008; Cinquini and Tenucci, 2010), the relevance of customer as an accounting object (McManus and Guilding, 2008).

Even if SMA is still considered in a development stage, needing a wider and clearer application in practice (Nixon and Burns, 2012), it represents a useful and potential bridge between MA and SDL. In this sense some of the SMA tools, like for instance Customer Profitability Analysis (CPA), OBA, SCM, IOCM, could assume a relevant role in the SDL context.

Grounding on the analysis of the previous sections, we can now identify six *critical aspects* or *dimensions* of business service-related change, in which it is particularly evident the challenge to MA research beyond most of the traditional approaches taken so far:

- Customer interaction, that focuses the linkage between the company and its customers;
- Profitability driver, as the determinants of company profitability;
- Measurement orientation, in term of relevant objects of company measurement process;
- Cost drivers, as the determinants of company costs;
- Resource position, or the role of integrated resources in the process of delivering value to customer;
- Revenues and price setting, considering the change in value creation and the new approaches required in setting prices.

Each of these topics requires new foundations in MA research that may represent a change of paradigm with respect to that of GDL that has prevailed in MA so far.

The following Table 4.1 summarizes the critical aspects characterizing management accounting and the approach adopted under a GDL as opposed to a SDL considering these issues.

The first relevant transition to SDL in management accounting comes from the *change in focus of customer interactions from a transaction-based to a relationship-based focus*. This can be observed as one of the most important consequences of the process of "servitization" in manufacturing. This shift implies the orientation towards a long-term relation with the customer and the transformation of the producer in a "service provider": an example is given by the development of service offering given by an equipment manufacturer (Oliva and Kallenberg, 2003). The implications in term

Table 4.1 Critical aspects for Management Accounting and their consideration under Goods-Dominant Logic (GDL) and Service-Dominant Logic (SDL).

Critical Aspect	GDL	SDL
Customer interaction (Value creation)	Transactions based (value in exchange)	Relationship-based (value in use)
Profitability driver	Minimize resource consumption (focus on efficiency)	Maximize resource usage (focus on capacity)
Measurement orientation	Product centric	Customer centric
Cost drivers	Volume-related	Capacity/Customer-related
Resource position	Resources owned to produce and sell "output" to customer (company measures)	Resources made available and integrated with partners to support customer in value co-creation process (network + customers measures)
Revenue and price setting	Production process driven (Cost-plus and Market)	Customer value co-creation driven (Capacity choice and Business model)

of accounting information required to support the process of servitization consist in different aspects. An interesting example is the case of Fleet management in HILTI (Bertini and Gourville, 2012). The company provides leading-edge technology to the global construction industry and decided to add the provision of a service (another example of servitizing company). Customers subscribing to Hilti's Fleet Management program pay a monthly fee that covers the use, service and repair of all tools. In this way customers are free from managing the tools and can concentrate on the business recognizing a sort of intangible value in the relationship with the company, such as the enhanced professionalism of using a set of high quality tools. Focusing on the relationships more than the transactions HILTI is able to create additional value for the customers.

Regarding *profitability driver,* the case of servitization shows how resource usage (capacity) is more relevant than resource consumption (cost). According with Oliva and Kallenberg (2003: p.168): *"The move towards maintenance contracts is often triggered by a desire to make better use of the installed service organization. For the service provider, once the service organization is in place, it becomes a fixed cost and the main driver of profitability is capacity utilization. Established service contracts reduce the variability and unpredictability of the demand over the installed capacity, and allow a higher average capacity utilization."*

It has been observed that the emerging service culture, with respect to the metrics, values and incentives predominant in the manufacturing organization, can be supported by an appropriate information system to monitor the business operations related to the servitization process, in order to demonstrate the contribution to profitability of the service organization activities within manufacturing (Oliva and Kallenberg, 2003). In this way also in web services (i.e. Google) the profitability is strictly linked to the capacity usage; the marginal cost of an additional unit of web-based service (i.e. Gmail service) is nearly zero and it provides a potential revenue becoming entirely profit (i.e. by the advertisements to be displayed on that new Gmail account). The fundamental choice for a company becomes capacity. Adopting a "customer-centric" thinking implies a detailed understanding of the activities a customer performs in using and operating a product through its life cycle, from sale to decommissioning (Davies, 2004). And coherently, also the *measurement orientation* needs a shift from the "cost of production" to the "cost of use" as part of its overall life cycle. The analysis of the costs of the services offered and their maintenance over time on the customer side assumes paramount relevance in the analysis. In this way the company is able to support strategies for innovative services linking the costs incurred (or to be incurred) and the utility by the user/customer. Costing techniques like Total Cost of Ownership or Life Cycle Costing greater fit such context.

Another important issue requiring a shift from GDL to SDL in management accounting is related to *cost driver*. In the most traditional GDL management accounting the core cost driver is still the production volume; the higher is the level of units sold and the higher is the level of costs. The main reason of cost incurred is the number of unit produced and sold. In SDL management accounting the ultimate cost driver is the capacity choice and the level of capacity utilization (both structural and operational cost driver of Riley (1987) classification). On this point, Time-driven Activity-based Costing (TDABC), a recent development of ABC in costing technique, goes further on the need of considering time as the main driver of capacity information. TDABC technique (Kaplan and Anderson, 2007) encounters the estimate of the practical capacity of committed resources, mainly people, and clearly fits the emergent need of service companies to better take into account the level of capacity usage.

Another differentiating aspect is the *resource position*: for GDL Management Accounting the resource consumption is crucial in order to attain a certain level of efficiency and being able to satisfy the customers through the acquisition of a product. The customer is satisfied only when he/she owns the product and resources are consequently used in that way. Under a SDL the company makes available the resources to the customer in order to increase his/her involvement in the value co-creation process. Moving the focus of value creation from exchange to use, means transforming our understanding of value from one based on units of firm output to one based on processes that *integrate resources* (Vargo et al., 2008). Under this perspective sharing resources with customers or other systems is critical. For instance, the recent *cloud computing* (more broadly *cloud sourcing*) phenomena, referring to the fact that the software is not downloaded but used on the web, is an example (i.e. Google Drive). The firm is fundamentally a value facilitator, but during interactions with its customers the firm may in addition become a co-creator of value with its customers. Firms produce input resources into customers' value-generating processes, and hence firms facilitate value creation providing indirect support to value creation. Such resources do not include value themselves. During interactions with customers firms get opportunities to influence their customers' value-generating processes and thus can become co-creators of value with their customers (direct support to value creation) (Gronroos and Ravald, 2009). Producer–consumer distinction is inappropriate and value is created through the active participation of all people/systems engaged in exchange. In this respect the before-mentioned model proposed by Gronroos and Helle (2010) represents an important reference to broaden measurement up to the value co-creation process shedding light on its economic evaluation side. Furthermore, value depends on the capabilities a system has to survive and accomplish other goals in its environment, and taking advantage of the service another system offers by *incorporating improved capabilities*. This is a process of utmost importance in generating value-in-use. It requires a renewed focus and new metrics: dimensions of system improvements in term of *resilience* (Dalziell and McManus, 2004; Whitson and Ramirez-Marquez, 2009), *adaptability* (Tuominen et al. 2004, Reeves and Deimler, 2011), *viability* (Fawcett et al., 2009) by means of resources integration within a particular environment are going to become challenging objects of measurement to Management Accounting.

A last but not less crucial field differentiating GDL to SDL management accounting is that of *revenue and price setting*. In the first type of accounting the price, as the cost records, is closely link to the product; it is calculated by the encounter of two methods: cost-plus or market. In the first case to the production unit cost (variable or full) is added a mark-up; in the second method the price is fixed on the base of a comparison with competitors' prices (Schlissel and Chasin, 1991). Change provided by servitization also affects pricing: from a mark-up for labor and parts every time a service is provided, to a fixed price covering all services over an agreed period. Relationship-based services centered around the product normally take the form of maintenance contracts priced in terms of operational availability and response time in case of failure. In SDL accounting price is driven more by *business model* and *capacity choices* than from resource consumption. The first refers to the recalled attention-based or transaction-based business model under which the monetary exchange can be realized or not. The capacity issue is critical, especially in service, in relation to the forecast of capacity usage in order to determine a reliable cost and, consequently, price.

We have therefore identified a set of six topics that challenge MA and its effectiveness in providing valuable information for business, according with the SDL principles and we pinpointed the change in approach required by MA considering these issues shifting from GDL to SDL. Some of these issues have already been claimed in research and literature inside and outside MA domain, and there are relevant contributions to be considered in the stock of knowledge available to move ahead the research in this area. In the following, these main contributions are recognized in relations to the Critical aspects of Table 4.1 and further challenging issues for Management Accounting research are given.

Regarding *Customer interaction*, Grönroos and Helle (2010) have proposed an evaluation model in which the joint supplier-customer productivity is considered as a consequence of–the efficiency and effectiveness of the relationship. Also Storbacka and Nenonen (2009) suggested a measure of the "value capture" by discounting the future profits arising from the relationship with the customer, and argued that this value can be used as a proxy of value creation for shareholders. Further, will it be possible to inform more the customer (e.g. about a user's real-time costs of using a firm's offering and perhaps the costs of integrating with other resource)? Will the customer be the only relevant external actor?

On the topic of *Profitability driver* the issue of productivity as a multidimensional concept has been discussed and qualified as the resultant of the interaction among internal efficiency (cost effects), external efficiency (related to perceived quality and providing revenue effects), and capacity efficiency (that influences costs, but also revenues) (Grönroos and Ojasalo, 2004). In dealing with servitization, Visnjic and Van Looy (2011) highlighted important aspects of service performance in an industrial, product driven global equipment manufacturer. Regarding the aforementioned crucial issue of capacity, an important issue is: how to measure and predict capacity usage and costs in a modular/networked environment?

If considering *Measurement orientation* toward customer, this is perhaps the most acknowledged change in MA as regard measurement object and several techniques have been proposed, implemented and discussed (Storbacka and Nenonen, 2009; Roslender and Wilson, 2012). Here a rising question is: will the customer profitability analysis be the only relevant? Which role may play TCO-LCC in value co-creation measurements?

Cost drivers analysis presents recent development by the proposal of Time-Driven Activity-Based Costing as an upgrade of Activity-Based Costing, that also constitutes an important stock of knowledge: TDABC is a cost management technique that emphasizes the cost of unused resources by using time as the cost driver to allocate indirect costs and time equations to customize costs of product/customers (Kaplan and Anderson, 2007). It is important to assess TDABC in action in service environments, particularly in measuring cost differences in serving customers and cost of unused capacity in activities and processes (Dorn and Seiringer, 2012). Assess the effectiveness of TDABC in service to measure cost of differences in serving customers and cost of unused capacity constitutes a valuable research task for MA here.

Resource position refers to the relevance of the integration between internal and external resources (customers/systems) to improve performance. Here, in addition to Ma tools in networks and Supply Chains, new dimensions of performance like resilience, adaptability, viability (measures of systems reliability studied in different research areas: Whitson and Ramirez-Marquez, 2009) become relevant objects of measurement in SDL perspective. Developing measures on dimensions of performance such as system resiliency, adaptability, viability and use them properly in control systems becomes therefore crucial also in MA perspective.

Revenue and price topic allows highlighting the change in role of cost modeling for pricing, when business models develop according to SDL. The price of goods and services may be discounted, in some cases even to free (as attention based business model – e.g. Google and other web services), to attract significant market shares (essential in the business of network services e.g. for a browser or a search engine like Google) or price can be used as a tool to draw attention on the product by a consumer that has become basically indifferent (Bertini and Wathieu, 2010). Prices may be also fixed to satisfy social needs and attain "shared value" with the pricing policy (e.g. London 2012 Olympic Games – Bertini and Gourville, 2012). Here the transition from price for transaction to price for utilization, and the consequent changing role of cost modeling for pricing, are relevant issue for MA. Challenging questions are: what will we price? Which requirements of cost modeling for pricing in the shifting from "price for transaction" to "price for usage"?

4.5 CONCLUSIONS

In this Chapter we addressed some possible implications that the change of perspective provided by the pillars of Service evolution in business may offer to the Management Accounting approach in supporting decision making. In our opinion this work represents only one step of the broader reflection on the role of accounting in the new paradigm of service (Kerr, 2008). The accounting domain still needs further and more intense development in the direction of service in order to maintain a informative role in the decision making process.

In this attempt, we have highlighted how the aspects characterizing SDL change the approach of management accounting in depth and the role it can play in supporting decision making in companies. In particular, we depicted six critical aspects or dimensions of business service-related change in which the approach of the traditional GDL perspective differs from that of SDL, and the MA implication for objects and drivers to focus by effective measurements. In doing this, we have focused the research that represents an important stock of knowledge for MA to develop further in this field.

These dimensions point out areas of MA to evolve further by research and practices in the vein of SDL. As we recalled, we believe that the changes identified by the SDL may fit into the wider change/evolution of management accounting identified in the framework of the Strategic Management Accounting. There is no agreed definition of SMA (Langfield-Smith, 2008; Cinquini and Tenucci, 2010), but it is well recognized it has a clear "external orientation", to be interpreted as the importance given to accounting information about competitors, suppliers and customers with the aim of providing information useful to set business strategic positioning.

This stream of research in MA also constitutes a fundamental stock to evolve in the direction addressed by SDL. The possible interfaces and integration with marketing and customer accounting in particular, represent areas of SMA research mostly developed in recent years (Roslender and Wilson, 2012). In this respect, within SMA context SDL accounting assumes a significant and coherent role for its focus on the customer side. SDL emphasizes the role of customer as value co-creator and consequently SDL accounting is addressed, among the others, to the challenge of the measurement of the part of the value co-created by the customer.

A service orientation further calls for the opening of the boundaries of the company. In the near future the inclusion of the customers and the suppliers (i.e. the entire supply chain) within the potential object of analysis for MA should be supplemented with the further inclusion of the other stakeholders. This means a "total opening" of the company towards external actors and a continuous exchange of financial and non-financial information among them.

Under such scenario company attends and participates in the ongoing interaction among the different actors of the service ecosystem. As such interactions generate trails, companies have large amount of data. If they are able enough to manage, process and analyse such financial and non-financial, quantitative and non-quantitative data, they can exploit huge market opportunities. Here MA could further contribute as an element of the wider and "new" informative system labelled as "Analytics 3.0" by Davenport (2013). The article provides the example of a servitizing company further focused on analytics: Schneider Electric. The company originally manufactured iron, steel, and armaments, but nowadays it focuses primarily on energy management, including energy optimization, smartgrid management, and building automation. Its Advanced Distribution Management System now handles energy distribution in utility companies, as an example of a service provided by an originally manufacturing company. As such a system is further used to monitor and control network devices or to manage service outages, it collects a big amount of data and allows employees to use analytics to manage the state of the network.

The opening of boundaries of the company increases its impact when added to the issue of resource integration in SDL. Resource integration is the means through which co-creation of value is achieved, and refers to the way the resources possessed by both the customer and supplier are utilized at the point of interaction. The merging of these two elements opens unexpected opportunities for potential new and unexplored performance measures such as resilience, adaptability and viability as they become potentially relevant objects of measurement.

Throughout the Chapter we have suggested some relevant issues and further critical areas that challenge service-oriented research in MA, namely the measurements of value in a relationship-based value creation, of new drivers of profitability, of new customer centric metrics, of capacity-related cost drivers and the role of costing, capacity and business modeling for pricing. It is our belief that all these issues may represent fruitful paths of research to embrace the acknowledgement of the impact of SDL in Management Accounting and to foster its innovation in the near future.

Table of notations

Notation	Definition
ABC	Activity-Based Costing
CPA	Customer Profitability Analysis
GDL	Goods-Dominant Logic
IOCM	Interorganizational Cost Management
LCC	Life-Cycle Costing
MA	Management Accounting
OBA	Open Book Accounting
OBC	Outcome Based Contracting
PBC	Performance Based Contracting
SCM	Strategic Cost Management
SDL	Service-Dominant Logic
SMA	Strategic Management Accounting
TCO	Total Cost of Ownership
TDABC	Time-Driven Activity-Based Costing
VCA	Value Creating Activities

REFERENCES

Barontini, R., Cinquini, L., Giannetti, R. and Tenucci, A. (2013) "Models of performance and value measurement in service systems", Cinquini, L., Di Minin, A. and Varaldo, R. (Eds), *New Business Models and Value Creation: A Service Science Perspective*, Springer, Milan.

Bertini M. and Wathieu L. (2010) "How to Stop Customers from Fixating on Price", *Harvard Business Review*, May, pp. 84–91.

Bertini M. and Gourville J.T. (2012) "Pricing to create shared value", *Harvard Business Review*, June, pp. 96–104.

Bhimani A. (2006) "Management Accounting and digitization", Bhimani, A. (Ed.), *Contemporary Issues in Management Accounting*, Oxford University Press, Oxford.

Bhimani A. and Bromwich M. (2010) *Management Accounting: Retrospect and prospect*. CIMA Publishing.

Brax, S. (2005) "A manufacturer becoming service provider – challenges and a paradox", *Managing Service Quality*, Vol. 15, No. 2, pp. 142–155.

Brimson J.A. and Antos J. (1994) *Activity-based Management for Service Industries, Government entities and Nonprofit organizations*, Wiley & Sons, New York.

Bromwich, M. (1990) "The case for strategic management accounting: the role of accounting information for strategy in competitive markets", *Accounting, Organizations and Society*, Vol. 15, pp. 27–46.

Burchell, S., Clubb, C., Hopwood, A., Hughes, J. and Nahapiet, J. (1980) "The roles of accounting in organizations and society", *Accounting, Organizations and Society*, Vol. 5, No. 1, pp. 5–27.

Caglio, A. and Ditillo, A. (2008) "A review and discussion of management control in inter-firm relationships: Achievements and future directions", *Accounting, Organizations and Society*, Vol. 33, No. 7, pp. 865–898.

Carr, C. and Ng, J. (1995) "Total cost control: Nissan and its U.K. supplier partnerships", *Management Accounting Research*, Vol. 6, pp. 347–365.

Chapman, C.S. (1997) "Reflections on a contingent view on accounting", *Accounting, Organizations and Society*, Vol. 22, No. 2, pp. 189–205.

Chen, X. (2011) "Interorganizational cost management in supply chain based on open book accounting", *The 2011 International Conference on Management and Service Science (MASS)*, Wuhan, China, 12–14 August 2011.

Cinquini, L. and Tenucci, A. (2010) "Strategic management accounting and business strategy: a loose coupling", *Journal of Accounting & Organizational Change*, Vol. 68, No. 2, pp. 228–259.

Coller, G., Tenucci, A. and Cinquini, L. (2011) Management accounting in Web services: Issues and challenges for a research agenda, paper presented at the EAA Annual Congress, Rome, 20–23 April 2011.

Cooper, R., Slagmulder, R. (1999) *Supply chain development for the lean enterprise. Interorganizationa cost management*, Productivity press, Portland, Oregon.

Cugini, A., Carù, A. and Zerbini, F. (2007) "The Cost of Customer Satisfaction: A Framework for Strategic Cost Management in Service Industries", *European Accounting Review*, Vol. 16, pp. 499–530.

Dalziell, E.P., McManus S.T. (2004) "Resilience, Vulnerability, and Adaptive Capacity: Implications for System Performance", *Presented at the International Forum for Engineering Decision Making (IFED)*, Stoos, Switzerland, December 6–8.

Davenport, T.H. (2013) "Analytics 3.0", *Harvard Business Review*, December, pp. 64–72.

Davies, A. (2004) "Moving base into high-value integrated solutions: a value stream approach", *Industrial and Corporate Change*, Vol. 13, No. 5, pp. 727–756.

Dearden, J. (1978) "Cost Accounting comes to service industries", *Harvard Business Review*, September–October, pp. 132–140.

Dekker, H.C. (2003) "Value chain analysis in interfirm relationships: a field study", *Management Accounting Research*, Vol. 14, pp. 1–23.

Dekker, H.C. and Van Goor, A.R. (2000) "Supply chain management and Management accounting: a case study of Activity-based Costing", *International Journal of Logistics: Research and Applications*, Vol. 3, No. 1, pp. 41–52.

Dorn, J., and W. Seiringer (2013) "A Prototype for Service-Based Costing", *Proceedings of the Annual Hawaii International Conference on System Sciences*, art. no. 6479993, pp. 1300–1309.

Drury, C. (2012) *Management and Cost Accounting*, Cengage Learning.

Ellram, L.M. (1993) "Total Cost of Ownership: Elements and Implementation", *International Journal of Purchasing and Materials Management*, Vol. 29, No. 4, pp. 3–10.

Ellram, L.M. and Siferd, S.P. (1998) "Total Cost of Ownership: a key concept in Strategic Cost Management Decisions", *Journal of Business Logistics*, Vol. 19, No. 1, pp. 55–84.

Fawcett, S.E., Allred C., Magnan G.M., Ogden J. (2009), "Benchmarking the viability of SCM for entrepreneurial business model design", *Benchmarking: An International Journal*, Vol. 16, pp. 5–29.

Garrison, R.H, Noreen, W. and Brewer P.C. (2008), *Managerial Accounting*, McGraw-Hill.

Gebauer, H. and Friedli T. (2005) "Behavioral implications of the transition process from products to services", *The Journal of Business & Industrial Marketing*, Vol. 20, No. 2, pp. 77–78.

Grönroos, C. (2000) *Service Management and Marketing: A Customer Relationship Management Approach*, John Wiley & Sons, Chichester.

Grönroos, C. (2008) "Service logic revisited: who creates value? And who co-creates?", *European Business Review*, Vol. 20, No. 4, pp. 298–314.

Grönroos, C. and Helle, P. (2010) "Adopting a service logic in manufacturing. Conceptual foundation and metrics for mutual value creation", *Journal of Service Management*, Vol. 21, No. 5, pp. 564–590.

Grönroos, C. and Ojasalo K. (2004) "Service productivity Towards a conceptualization of the transformation of inputs into economic results in services", *Journal of Business Research*, Vol. 57, pp. 414–423.

Grönroos, C. and Ravald, A. (2009) "Marketing and the logic of service: value facilitation, value creation and co-creation and their marketing implications", Working Paper 542, Hanken School of Economics, Helsinki.

McManus, L. and Guilding, C. (2008) "Exploring the potential of customer accounting: a synthesis of the accounting and marketing literatures", *Journal of Marketing Management*, Vol. 24, No. 7–8, pp. 771–795.

Hilker, C. (2013) "User Interaction Revolution", *Cost Management*, Jul/Aug, online article.

Hoffjan, A. and Kruse, H. (2006) "Open book accounting in supply chains – When and how is it used in practice?", Journal of Cost Management, Vol. 20, No. 6, pp. 40–47.

Hopwood, A.G. (1980) "Organizational and behavioural aspects of budgeting and control", Arnold, J., Carsberg, B. and Scapens, R. (Eds), *Topics in Management Accounting*, Philip Allan Publishers, Oxford.

Horngren, C.T., Bhimani, A., Datar, S.M., Foster, G. (2001) *Management and Cost Accounting*, PrenticeHall.

IBM (2004) IBM Research. Service Science. A New Academic Discipline?. Paper downloaded at: http://www.almaden.ibm.com/asr/SSME/

IBM and IfM (2008) *Succeeding through service innovation. A service perspective for education, research, business and government*, University of Cambridge, Institute for Manufacturing, Cambridge (UK).

Kajuter, P. and Kulmala, H. (2005) "Open-book accounting in networks. Potential achievements and reasons for failures", *Management Accounting Research*, Vol. 16, No. 2, pp. 179–204.

Kaplan, R.S. and Anderson, S.R. (2007) *Time-Driven Activity-Based Costing*, Boston, Harvard Business School Press.

Kaplan R.S. and Norton, D.P. (2008) *Execution Premium*, Harvard Business School Press.

Kerr, S. G. (2008) "Service Science and Accounting", *Journal of Service Science* 1, 17–26.

Langfield-Smith, K. (2008) "Strategic management accounting: how far have we come in 25 years?", *Accounting, Auditing and Accountability Journal*, Vol. 21, No. 2, pp. 204–228.

Laine, T. (2009) *Exploring Pilot Projects of a Manufacturer on Service R&D to Understand Service as an Accounting Object*, Tampere University of Technology.

Laine, T., Paranko, J. and Suomala, P. (2012) "Management accounting roles in supporting servitisation. Implications for decision making at multiple levels", *Managing Service Quality*, Vol. 22, No. 3, pp. 212–232

Lev, B. (2001) *Intangibles: Intangibles: Management, Measurement, and Reporting*, Brookings Institution Press, Washington.

Lin, Y., Shi, Y. and Zhou, L. (2010) "Service Supply Chain: Nature, Evolution, and Operational Implications." in *Proceedings of the 6th CIRP-Sponsored International Conference on Digital Enterprise Technology*, Vol. 66, 1189–1204 (Eds G. Huang, K. L. Mak and P. Maropoulos). Springer Berlin Heidelberg.

Maglio P. and Spohrer, J. (2008) "Fundamentals of service science", *Journal of the Academy of Marketing Science*, Vol. 36, No. 1, pp. 18–20.

Mathieu, V. (2001) "Service strategies within the manufacturing sector: benefits, costs and partnership", *International Journal of Service Industry Management*, Vol. 12, No. 5, pp. 451–475.

Modell S. (1996) "Management accounting and control in services: structural and behavioural perspectives", *International Journal of Service Industry Management*, Vol. 7, No. 2, pp. 57–80.

Mouritsen, I., Hansen, A. and Hansen C. (2001) "Inter-organizational controls and organizational competencies: episodes around target cost management/functional analysis and open book accounting", *Managment Accounting Research*, Vol. 12, pp. 221–244.

Neely, A. (2008) "Exploring the financial consequences of the servitization of manufacturing", *Operations Management Research*, Vol. 1, No. 2, pp. 103–118.

Neely, A., Benedetinni, O. and Visnjic, I., (2011) "The servitization of manufacturing: Further evidence", Academic paper presented at the 18th European Operations Management Association Conference, Cambridge, July 2011.

Ng, I. (2010) "The future of pricing and revenue models", *Journal of Revenue and Pricing Management*, Vol. 9, pp. 276–281.

Ng, I., Nudurupati S.S., Tasker P., T. (2010) "Value co-creation in the delivery of Outcome Based Contracts for Business-to-Business service", *AIM-Research Working Paper Series*, Cranfield (UK).

Ng, I., Parry, G., Smith, L., Maull, R. and Briscoe, G. (2012) "Transitioning from a goods-dominant to a service-dominant logic. Visualising the value proposition of Rolls-Royce", *Journal of Service Management*, Vol. 23, No. 3, pp. 416–439.

Normann, R. (2001) *Reframing Business: When the Map Changes the Landscape*, Chichester, Toronto, Wiley.

Normann, R. and Ramirez, R. (1993) "From value chain to value constellation: Designing interactive strategy", *Harvard Business Review*, Vol. 71, No. 4, p. 65

Oliva, R. and Kallenberg, R. (2003) "Managing the transition from products to services", *International Journal of Service Industry Management*, Vol. 14, No. 2, pp. 160–172.

Ostrom, A.L., Bitner, M.J., Brown, S.W., Burkhard, K.A., Goul, M., Smith-Daniels, V., Demirkan, H. and Rabinovich, E., 2010. "Moving forward and making a difference: Research priorities for the science of service", *Journal of Service Research*, Vol. 13, No. 1, pp. 4–36.

Randall, W. S., Pohlen, T. L. and Hanna, J. B. (2010) "Evolving a theory of Performance-based Logistics using insights from Service Dominant Logic", *Journal of Business Logistics*, Vol. 31, pp. 35–61.

Ravald, A. and Gronroos, C. (1996) "The value concept in marketing", *European Journal of Marketing*, Vol. 30, No. 2, pp. 19–30.

Reeves, M., Deimler M. (2011) "Adaptability: The New Competitive Advantage", *Harvard Business Review*, July–August, pp. 135–141.

Riley, D. (1987) "Competitive cost based investment strategies for industrial companies", Booz, Allen and Hamilton (Eds.), *Manufacturing Issues*, New York.

Roslender, R. and Wilson, R.M.S. (2012) *The Marketing/Accounting Interface*, Routledge.

Schlissel, M. R. and Chasin J. (1991) "Pricing of Services: An Interdisciplinary Review", *Service Industries Journal*, Vol. 11, No. 3, pp. 271–286.

Schwab, K. (Ed.) (2010) *The Global Competitiveness Report 2010–2011*, The World Economic Forum, Geneva.

Shank, J.K. and Govindarajan, V. (1993) *Strategic Cost Management: The New Tool for Competitive Advantage*, The Free Press, New York, NY.

Shostack, G. L. (1977) "Breaking free from product marketing", *Journal of Marketing*, Vol. 41, No. 2, pp. 73–80.

Slack, N. (2005) *Patterns of Servitization: Beyond Products and Service*. Cambridge: Institute for Manufacturing, Cambridge University.

Spohrer, J. and Kwan, S. K. (2009) "Service Science, Management, Engineering, and Design (SSMED): An Emerging Discipline – Outline & References", *International Journal of Information Systems in the Service Sector*, Vol. 1, No. 3, pp. 1–31.

Spohrer, J., Kwan S.K., Demirkan H. (2013) "Service science: on reflection", Cinquini L., Di Minin A., Varaldo R. (Eds.), *New Business Models and Value Creation: A Service Science Perspective*, Milano, Springer.

Stewart, T.A. (1997), *Intellectual Capital; The New Wealth of Nations*, Doubleday, New York.

Storbacka, K. and Nenonen, S. (2009) "Customer relationships and the heterogeneity of firm performance", *Journal of Business and Industrial Marketing*, Vol. 24, No. 5/6, pp. 360–372.

Tuominen, M., Rajala A., Möller K. (2004) "How does adaptability drive firm innovativeness?", *Journal of Business Research*, Vol. 57, pp. 495–506.

Vandermerwe, S. and Rada, J. (1988) "Servitization of business", *European Management Journal*, Vol. 6, No. 4, pp. 314–324.

Vargo, S.L. and Akaka, M. (2009) "Service-Dominant Logic as a Foundation for Service Science: Clarifications", *Service Science*, Vol. 1, No. 1, pp. 32–41.

Vargo, S.L. and Akaka, M. (2012) "Value Cocreation and Service Systems (Re)Formation: A Service Ecosystems View", *Service Science*, Vol. 4, No. 3, pp. 207–217.

Vargo S.L. and Lusch R.F. (2004) "Evolving to a new dominant logic for Marketing", *Journal of Marketing*, Vol. 68, No. 1, pp. 1–17.

Vargo, S. L., and Lusch, R. F. (2006) "Service-dominant logic: What it is, what it is not, what it might be", Vargo S.L. and Lusch R.F. (Eds.), The service-dominant logic of marketing: Dialog, debate, and directions, (pp. 43–56). Armonk, NY: ME Sharpe.

Vargo, S.L. and Lusch R.F. (2008) "Service-dominant logic: continuing the evolution", *Journal of the Academy of Marketing Science*, Vol. 36, pp. 1–10.

Vargo, S.L., Maglio, P.P., Akaka M.A. (2008) "On value and value co-creation: A service systems and service logic perspective", *European Management Journal*, Vol. 26, pp. 145–152.

Visnjic, I. and Van Looy, B. (2011) "Can Manufacturers Become Service Providers? Fostering Complementarity between Products and Service", paper presented in a Divisional Roundtable Paper Session, Academy of Management Annual Meeting, 2011.

Visnjic, I. and Neely, A., Wiengarten, F., (2012) "Another performance paradox? A refined view on the performance impact of servitization", Working Paper, May, University of Cambridge.

Whitson, J.C., Ramirez-Marquez J.E. (2009), "Resiliency as a component importance measure in network reliability", *Reliability Engineering & System Safety*, Vol. 94, pp. 1685–1693.

Yoshikawa, T., Innes, J. and Mitchell, F., (1993) Japanese Cost Management Practices, in Brinker B. (Ed.), *Handbook of Cost Management*, Warren, Gorham & Lamont, Boston.

Chapter 5

Managing service supply chains in the big data era: A system of systems perspective

Tsan-Ming Choi

Business Division, Institute of Textiles and Clothing, The Hong Kong Polytechnic University, Hung Hom, Kowloon, Hong Kong

SUMMARY

Supply chain management is one of the most important areas in modern business. In the domain of services operations management, service supply chains are multi-echelon dynamic systems which focus on how collaboration and coordination among supply chain agents can enhance service management and achieve global optimality. Undoubtedly, service supply chain management has become an especially important topic in the big data era, in which the massive amount of data would provide great opportunity to advance it. This chapter discusses, from the system of systems perspective, how the service supply chain can be better managed. We first present the characteristics of service supply chain systems. Then, we discuss why the service supply chain is qualified to be called a system of systems. Afterwards, we highlight some technological elements related to service supply chains and discuss some principles from the system of systems perspective to enhance service supply chain management. Future research directions are proposed.

Keywords

Service supply chain management, system of systems, service science, operations management, big data, systems engineering

5.1 INTRODUCTION

With the advance of information technology, the concept of big data arises. The common characteristics of "big data" include: (i) the size of the dataset is huge, and (ii) the available computing facilities are insufficient to handle such a massive amount of data. Undoubtedly, nowadays, from social media, e-commerce websites, digital customer profiles and transactions records, service companies can easily obtain "big data" relevant to their operations. For example, a financial bank, which offers e-banking services, can keep track of the customer behaviours online and provide more customized services. A hospital is able to obtain the big data regarding its patients and all kinds of data would be useful for healthcare suppliers and drug suppliers. A cable TV provider can observe from the "on-demand" services what customers like and fine-tune the future offering. It is rather widely believed that big data would revolutionize many management practices (see McAfee and Brynjolfsson (2012) for more discussions).

Undoubtedly, we are now in the big data era, and the respective service supply chains exhibit the following characteristics:

1   Services play a critical role in any service supply chains. This is a natural feature as the respective supply chains are "service supply chains". Popular examples of "pure service supply chains" can be found in the service industries such as design, fashion beauty, finance, healthcare, media, etc. As a remark, we do not rule out the "product-service supply chains" in which there are both physical products and services (see Wang et al. (2015) for more discussions on these two kinds of service supply chains).

2   Each agent of the service supply chain can be an independent entity by itself and it has its own functionality. As an example, in finance, the retail bank provides direct services to consumers and industrial customers (e.g., other companies). The upstream suppliers of information systems provide the necessary computing support to the banking service operations for both online (e.g., e-banking) and offline channels. There are service providers on intelligent business analytics to provide solutions to help companies make a wise use of the collected "big data". There are also many other partners of the bank such as credit card companies and the related governing parties. All of these agents work together to support this banking financial service supply chain and each agent has its own functionality.

3   The service supply chain is highly dynamic and will evolve with respect to the market needs. Four decades ago in the 1980s, e-banking was just something in people's dream and the banking financial service supply chain at that time did not offer any important service online. When time passes, nowadays, e-banking is commonly available in well-established banks all around the world. From e-banking and other e-commerce activities, agents of the financial service supply chains can make use of the collected huge datasets to make intelligent business decisions, such as providing the tailor-made offer on financial services to each individual customer. This example illustrates the dynamic and evolving nature of service supply chains. In fact, when time passes, the market requirements and the environment both change and it is just natural for the service supply chain to evolve and adapt to the new environment.

4   The whole service supply chain serves the respective market need which is not something achievable by each individual supply chain agent alone. This feature is very intuitive. If we take a look at the banking financial service supply chain we mentioned in the previous points, the services provided by the whole service supply chain cannot be provided by any one of the supply chain agents alone.

5   In general, there are multiple agents of the service supply chains. They are usually located in different geographical places and locations. For example, when we look at the service agents in the retail operations, no matter we talk about the financial service providers such as banks or fashion beauty salons, we can easily find that there are upstream suppliers supporting these retail operations. Thus, all these service supply chains involve multiple agents and in most cases, the agents are located in different places. In the big data era, this point also applies to information systems service providers such as the cloud application services and web server

systems, in which each service provider can be treated as an agent of the service supply chain.

5.2   SYSTEM OF SYSTEMS

In systems engineering, a system is usually interpreted as a collection of components organized together to help achieve a common goal or function. This goal or function is not going to be achievable by each component or a subset of components alone (Maier 1998). The system of systems concept has been well-discussed for decades (see the review by Gorod et al. 2008) and a system of systems can be described as *"a set of arrangement of interdependent systems that are related or connected to provide a given capability"* (Khalil et al. 2012). It is usually associated with a large scale and complex hierarchical structure, and possessing multiple objectives (Haimes and Li 1988). The focal point of a system of systems is on collaboration among the component systems so that new functionality and performance can be created (Baldwin et al. 2012). We list the key identifying characteristics of a system of systems in the following (Maier 1998, Sage and Cuppan 2001, Keating et al. 2003, Madni and Sievers 2014):

1   **Operational independence:** This independence refers to the component systems. If the system is qualified to be a system of systems, then its component systems must be able to operate independently by themselves.
2   **Managerial independence:** This independence implies that the component systems of the system of systems not only can operate independently, they indeed do operate and function as self-contained systems independently.
3   **Evolutionary nature:** The system of systems is not static and fully formed. It will evolve when time passes in which its functionalities will be revised from time to time and new purposes will also be added.
4   **Emergent behavior:** The system of systems achieves the objectives and performs the functions not achievable by its component systems.
5   **Geographic distribution:** The component systems are usually geographically dispersed as the system of systems is usually very complex and big in size.

Notice that points 1 and 2 are proposed to be the most important and critical characteristics by Maier (1998) and he also argues that the other characteristics are less important, in particular, the geographic distribution characteristic should be removed in many cases.

Now, a natural question arises: Are service supply chain systems in the big data era qualified to be called systems of systems? The answer is crystal clear now: If we check the five characteristics of system of systems with the five points we raised in Section 5.1, we will find that service supply chains fully satisfy the five common characteristics of systems of systems and hence we can confidently conclude that the service supply chains belong to the class of systems of systems. Table 5.1 summarizes this finding. As a consequence, the technologies which can enhance the development of systems of systems and the principles which are applicable to support systems of systems would all be applicable to service supply chain management.

Table 5.1 Service supply chains as systems of systems.

Characteristics	Service Supply Chains	Satisfied?
Operational independence	Individual service supply chain agents can operate independently	Yes
Managerial independence	Individual service supply chain agents do operate independently	Yes
Evolutionary nature	Service supply chains are evolving when time passes	Yes
Emergent behavior	Service supply chains achieve the functions not achievable by individual agents alone	Yes
Geographic distribution	Agents of the service supply chain are usually geographically dispersed	Yes

5.3   IMPORTANT TECHNOLOGIES FOR SERVICE SUPPLY CHAINS OPERATIONS

In Section 5.2, we have established that the service supply chain is in fact a system of systems. To support the service supply chain in the big data era, as a system of systems, some important technologies are widely applied. In this section, we examine them one by one.

The first technology is the cloud computing technology (Liu 2013, Khurana 2014), which was defined by the National Institute of Standards and Technology in 2011 (see Haimes et al. 2015) as "*a model for enabling convenient, on-demand network access to a shared pool of configurable computing resources (e.g., networks, servers, storage, applications, and services) that can be rapidly provisioned and released with minimal management effort or service-provider interaction*". There is no doubt that the cloud computing technology provides flexibility and agility to support supply chain operations for the service industry in the big data era as all the named computing facilities and resources are valuable and critical to the success of the respective service operations. Notice that even though the cloud computing technology has its niche and appealing features, it is well argued that users of the cloud computing technology will face a higher level of risk compared to the users who do not employ the cloud computing technology. Thus, risk management becomes a top agenda for service supply chains which employ the cloud computing technology (Haimes et al. 2015).

The second technology is mobile computing (Sefid-Dashti and Habibi 2014). With the advance of mobile computing technologies, smart phones and tablet devices are commonly used by almost everybody in the world (at least in the developed countries). From the service supply chain perspective, it is critical to take advantage of this technology in several ways. First, for the market side, consumers are using mobile devices and hence many services provided by the service supply chain have to be compatible with these mobile devices. Second, from the mobile commerce platform and transactions, big data can be collected and service supply chains should make a good use of the collected data to improve the service offering and improve profitability by proper revenue management (e.g., adopting a dynamic pricing strategy and offering customized services to individual consumers). Third, staff members of the service supply chains can employ mobile technologies to facilitate their own job tasks and operations.

The other critical technologies include data mining (for generating decision supports and insights from the collected "big data") and the information exchange and communication platform (for supporting the collaborative service supply chain operations and synchronizing the activities in the service supply chain system of systems). Since the related technologies in the big data era have been well-examined, we refer readers to the related studies for more details (Choi et al. 2016a).

5.4 KEY PRINCIPLES

After identifying the supporting technologies for service supply chain operations, in this section, we propose some principles for supporting the sustainable development of service supply chains from the perspective of system of systems. Observe that these principles are adapted and refined from the related literature (Rechtin 1991, Maier 1998, Volkert et al. 2014, Choi et al. 2016a and 2016b).

1   **Proper risk management to establish stability:** Rechtin (1991) proposes that it is important to establish stability in the system of systems. He argues that a system of systems will develop more promptly if stability is well-established. From the system of systems perspective, a stable form of the service supply chain is self-supporting technically and economically. Here, the technical stability refers to the condition in which the service supply chain can operate and deliver what it aims to achieve. The economic stability means that the service supply chain is economically self-sustainable. It is obvious that most service supply chains are open to threats from, e.g., the market demand volatility. To achieve a stable system of systems requires agents of the service supply chain to collaboratively develop a robust risk management scheme (Choi et al. 2016b).

2   **Emphasizing on cooperation and coordination:** One central theme in modern supply chain management is on supply chain coordination. A supply chain is coordinated when its agents make a decision which is also the best for the whole supply chain. Undoubtedly, to achieve coordination, it requires imposing the right incentives on the related parties and also encouraging faithful cooperation and collaboration among supply chain agents. This point is in line with Maier (1998)'s principle for system of systems on "*Ensuring Cooperation*" in which he proposes that "*if a system requires voluntary collaboration, the mechanism and incentives for that collaboration must be designed in.*" In service supply chains, incentive alignment contracts and the respective mechanisms can be applied so that individual agents will work towards the optimal solution for the whole service supply chain willingly (see Wang et al. 2015 and Choi et al. 2016b for more related studies of service supply chain coordination). As a remark, to support these supply chain coordination schemes, information technologies for proper information exchange are crucial or else moral hazards (like cheating) will very likely occur.

3   **Adopting the policy triage:** In the service supply chain, there are many agents. Some agents are weaker, some are performing normally, and some are great achievers.

By the policy triage proposed by Rechtin (1991) and echoed by Maier (1998), we should "*let the dying die. Ignore those who will recover on their own. And treat only those who would die without help*". In the service supply chain, it will be wise to focus our attention on and allocate resources to the needy in the supply chains. Thus, for the weakest links (i.e., agents) of the service supply chain which have no hope to "recover", we should remove them and find new ones. For the agents who can recover and need support, we should provide the needed assistance in a timely manner. Notice that this point is especially important for the selection of technological service providers as some of them simply cannot fulfil the task well and should hence be replaced. It is unwise to put further resources into them as the effort will go in vain in many cases.

4   **Focusing on the interfaces:** As what Maier (1998) proposes, the greatest leverage and danger in system of systems architecting are on the interfaces between component systems. In a service supply chain, these interfaces refer to how the supply chain agents interact and are usually governed by well-designed agreements and contracts. Adopting this principle, proper service supply chain management is built upon the appropriate management of these "interfaces". This point is in line with the above point on "Emphasizing on cooperation and coordination".

5   **Including uncertainty:** In supply chain management, traditional practices would include some technical performance measures to help monitor and predict if the planned achievement can be accomplished (Cho et al. 2012). In many cases, these technical performance measures are set in a deterministic way. However, service supply chains, as systems of systems, are complex and highly dynamic. Thus, it is important to incorporate uncertainty into the supply chain development performance measures. This whole scheme is critically important yet uneasy to implement (see Volkert et al. 2014 for more discussions).

6   **Employing the right information technologies:** Nowadays, service supply chain management requires an extensive use of technologies. In the big data era, information technologies will be applied from input, processing, to output stages of the service supply chain system of systems. Computing systems such as RFID technologies, e-commerce, and mobile apps, will be used for "inputs", cloud computing applications, and business analytics technology solutions will be used for "processing", and customized service offering is the usual output of a service supply chain. For all these steps, it is pertinent to understand that the key point is select the most appropriate technology, but not necessarily the most advanced technology, to deploy. Scalability (for future development), cost-and-benefit trade-offs, systems security, and the level of stability (with respect to the technology's level of maturity and the experience of service supply chain agents) of the associated technologies are some important selection criteria for determining the right information technologies to employ.

Table 5.2 summarizes the core insights/implications from the above proposed key principles. From Table 5.2, we can see that these key principles provide important rules of thumbs and guidance on how we can improve the management of service supply chain systems of systems.

Table 5.2 Key principles for managing service supply chain system of systems in big data era.

Key Principles	Core Insights/Implications
Proper risk management to establish stability	Implement robust collaborative risk management schemes to create stability in service supply chains
Emphasizing on cooperation and coordination	Proper use incentive supply contracts to achieve coordination and establish cooperative relationship
Adopting the policy triage	Help the service supply chain agent which can recover and remove the weak links
Focusing on the interfaces	Pay attention to interfaces (agreements and contracts between agents) helps develop solid service supply chain systems of systems
Including uncertainty	Incorporate uncertainty into service supply chain development performance measures
Employing the right information technologies	Identify the right, but not necessarily the latest, information technologies with full considerations of multiple factors

5.5 CONCLUSION

In this chapter, we have discussed service supply chain management in the big data era from the system of systems perspective. First, we have discussed what it means by big data and some features of service supply chains in the big data era. Then, we have presented the list of commonly adopted characteristics of the systems of systems and established that the service supply chain in the big data era is in fact a kind of system of systems. After that, we have examined a few enabling technologies for service operations. Finally, following the system of systems literature, we have proposed several key principles on proper service supply chain management in the big data era. Future research can be conducted in several areas. For example, one can study and validate the applicability of the key principles proposed in this paper in different real world service supply chains. One may also drill deeper and discuss the idea of "system of systems integration" (Madni and Sievers 2014) in service supply chains. Since risk management is also a pertinent issue in service supply chains, it will be interesting and promising to develop a methodology on risk management for service supply chains from the system of systems perspective.

REFERENCES

Arkoff, R.L. Toward a system of systems concept. *Management Science*, 17, 661–671, 1971.

Baldwin, W.C., T. Ben-Zvi, B.J. Sauser. Formation of collaborative system of systems through belonging choice mechanisms. *IEEE Transactions on Systems, Man and Cybernetics, Part A: Systems and Humans*, 42(4), 793–801, 2012.

Cho, D.W., Y.H. Lee, S.H. Ann, M.K. Hwang. A framework for measuring the performance of service supply chain management. *Computers & Industrial Engineering*, 62, 801–818, 2012.

Choi, T.M., H.K. Chan, X. Yue. Recent development in big data analytics for business operations and risk management. IEEE Transactions on Cybernetics, in press, 2016a.

Choi, T.M., Y. Wang, S.W. Wallace. Risk management and coordination in service supply chains: information, logistics and outsourcing. *Journal of the Operational Research Society*, 67, 159–164, 2016b.

Gorod, A., B. Sauser, J. Boardman. System-of-systems engineering management: review of modern history and a path forward. *IEEE Systems Journal*, 2(4), 484–499, 2008.

Haimes, Y.Y., B.M. Horowitz, Z. Guo, E. Andrijcic, J. Bogdanor. Assessing systemic risk to cloud-computing technology as complex interconnected systems of systems. *Systems Engineering*, 18(3), 284–299, 2015.

Haimes, Y.Y., D. Li. Hierarchical multiobjective analysis for large-scale systems: Review and current status. *Automatica*, 24(1), 53–69, 1988.

Keating, C., R. Rogers, R. Unal, D. Dryer, A. Sousa-Poza, R. Safford, W. Peterson, G. Rabadi. Systems of systems engineering. *Engineering Management Journal*, 15(3), 36–45, 2003.

Khalil, W., R. Merzouki, B. Ould-Bouamama, H. Haffaf. Hypergraph models for system of systems supervision design. *IEEE Transactions on Systems, Man and Cybernetics, Part A: Systems and Humans*, 42(4), 1005–1012, 2012.

Khurana, A. Bringing big data systems to the cloud. *IEEE Cloud Computing*, 1, 72–75, 2014.

Liu, H. Big data drives cloud adoption in enterprise. *IEEE Internet Computing*, 17, 68–71, 2013.

Madni, A.M., M. Sievers. System of systems integration: key considerations and challenges. *Systems Engineering*, 17(3), 330–347, 2014.

Maier M.W. Architecting principles for systems-of-systems. *Systems Engineering*, 1(4), 267–284, 1998.

McAfee, A., E. Brynjolfsson. Big data: the management revolution. *Harvard Business Review*, 90, 60–68, 2012.

Rechtin, E. *Systems Architecting: Creating and Building Complex Systems*. Prentice Hall, Englewoord Cliffs, NJ, 1991.

Sage, A.P., C.D. Cuppan. On the systems engineering and management of systems of systems ad federations of systems. *Information, Knowledge, Systems Management*, 2(4), 325–345, 2001.

Sefid-Dashti, B., J. Habibi. A reference architecture for mobile SOA. *Systems Engineering*, 17(4), 407–425, 2014.

Wang, Y., S.W. Wallace, B. Shen, T.M. Choi. Service supply chain management: A review on operational models. *European Journal of Operational Research*, 247, 685–698, 2015.

Volkert, R., J. Stracener, J. Yu. Incorporating a measure of uncertainty into systems of systems development performance measures. *Systems Engineering*, 17(3), 297–312, 2014.

Part III

Industrial applications and cases

Chapter 6

Evaluation of discrete event simulation software to design and assess service delivery processes

Giuditta Pezzotta, Alice Rondini, Fabiana Pirola & Roberto Pinto
CELS – Research group on Industrial Engineering, Logistics and Service Operations, DIGIP – Department of Management, Information and Production Engineering, Università degli Studi di Bergamo, viale Marconi 5, Dalmine (BG), Italy

SUMMARY

One of the major difficulties encountered by companies nowadays is related to the engineering of service and product-service systems (PSS), including the definition of methods and techniques supporting their design and development. It is noteworthy to state that, in order to make PSS and service provision profitable in the long term, it is of utmost relevance to balance the excellence in the value provided to the customer with a high efficiency and productivity of the service processes. In this respect, we posit that Discrete Event Simulation (DES) can effectively support managers in defining this equilibrium. Since the major simulation tools have been originally developed to simulate manufacturing processes, the purpose of this paper is to perform a comparative analysis between four simulation software packages, evaluating which one is the most suitable to be used in the PSS and service contexts. Thanks to the data collected in a real case (Take-a-way food shop with home-delivery) a comparison of the simulators, based on a selection of criteria from the literature and on specific features of service process, has been performed. The real case helped to shed the light on strengths and weaknesses of the analysed simulation software in supporting the design of a service process.

Keywords
Service Engineering, Service delivery process, Service Design, Simulation, Discrete events simulation, Software comparison

6.1 INTRODUCTION

The service sector has become the largest sector in most of the industrialized countries, and it continues to expand (Xiaofei & Wang, 2011). This is due to the critical role that services cover in current business and to the evolution of many manufacturing companies towards the provision of Service Environment (PSS) (Vandermerwe & Rada, 1988; Baines, et al., 2007).

In this context, the design and development of effective and efficient services and PSSs raise new issues, investments in numerous assets, processes, people, and materials, especially concerned with the peculiar requirements introduced by the thousands of components a service and a PSS are made of (Goldstein, et al., 2002).

Designing and developing service and PSS is a complex task, especially due to unpredictability of the solution lifecycle and to the number of interactions existing between the involved actors and the constituent components (Pezzotta, et al., 2012). To (partially) respond to these requirements, Service Engineering (SE) has emerged as a discipline explicitly addressing the design and the development of service and/or PSS offerings. In fact, SE is defined as a "technical discipline concerned with the systematic development and design of PSS using suitable models, methods, and tools" (Bullinger, et al., 2003). In SE, great attention must be given to the analysis of performance of the designed solution, from both an internal (company performance) and an external perspective (value for customer). In particular, the service delivery process is the most critical part of the solution, since it mainly affects the customer value (i.e. customers' satisfaction along different dimensions) and company performance (i.e. resources' usage and saturation, and costs). To this end, a process simulation approach can support decision makers in evaluating alternative solution from a dynamic perspective under different conditions and scenarios (Pezzotta, et al., 2014; Chalal, et al., 2015; Visintin, et al., 2014).

The purpose of this chapter is to provide a comparative analysis of available simulation software solutions and paradigms suitable for service delivery process simulation. We considered three commercial, rather general-purpose packages and one domain-specific package designed specifically to simulate a service with its distinctive features. The evaluation has been performed modelling a real service delivery process (in particular, a take-away food shop with home-delivery option). The software comparison has been carried out based on a selection of criteria gathered from the literature and a set of specific features of a pure service process.

To this end, the chapter is structured as follows: introduction on service design and simulation are reported in section 6.2 and section 6.3 respectively, while section 6.4 reports the methodology adopted to compare the software packages. The main considerations and results are discussed in section 6.5. Conclusions and further development are presented in section 6.6.

6.2   THE DESIGN OF SERVICES

The capability to innovate knowledge-intensive services is an order qualifier for almost all the existing service and manufacturing companies. Moreover, the ability to create business and societal values from expertise in varied domains and to design, develop and deliver new technology and services is considered one possible way to exploit a competitive advantage in the long-term (Quinn, et al., 1987; Marc, 2012). In this sense, a proper design can support companies in making their service useful, usable, efficient, effective and desirable (UK Design Council, 2010).

From a deep analysis of the service design literature (Heskett, 1987; Armistead, 1990; Goldstein et al., 2002; Roth & Menor, 2003), it results evident that in order to create a valuable service it is fundamental to align business strategy, service concept, and the design of the service delivery process.

In fact, a holistic approach involving marketing, human resources, operations, organizational structure, and technology disciplines is needed in order to orchestrate all the different service components (Ostrom et al., 2010). Therefore, it is necessary to model the right physical environment, people (customers and employees), and service delivery process to help customers to reach their desired experiences (Teixeira, et al., 2012).

However, the service components introduce design requirements that are profoundly different from those related to the product, thus necessitating an ad-hoc approach going beyond the mere transposition of product-design methodologies to the service domain.

In fact, unlike product, service components are a combination of processes, people skills, and materials that must be appropriately integrated to result in the 'planned' or 'designed' service (Goldstein, et al., 2002).

Particularly relevant to our discussion – and for the adoption of a simulation approach for analyzing the service delivery processes performance – is the possibility to consider the service delivery as a business process (Ponsignon, et al., 2012) composed of coordinated activities and tasks, performed by available resources, according to the customers' needs.

6.2.1   Service as a process

According to (Cook, et al., 1999), definitions of service and its role in the economy have been reported in literature since the early 1960s, but it seems that currently "there is no standardized definition of a service" (Parker, 2012). The US Department of Commerce's Standard Industrial Classification (SIC) defined as services all those activities that did not fit into agriculture or manufacturing, while the American Marketing Association (AMA) in the 1960s denoted service as: "Activities, benefits, or satisfactions which are offered for sale or are provided in connection with the sale of goods".

One of the most recent definitions, proposed by IBM in the Service Science research program, described service as "a provider/client interaction in which both parties participate and both parties obtain some benefit from the relationship [...]. A service is a form of activity, consumed at the point of production." (Katzan, 2011).

Analyzing these definitions, it is evident that services can be associated with "activities" or "processes" performed by the service provider. Accordingly, Berry (1980) and Zeithaml & Bitner (1996) defined service as "deeds, acts or performance", whereas Gronroos (2000) presented a service as "an activity or series of activities provided as a solution to customer problems" and Edwardsson, et al. (2005) stressed the process nature of a service and the benefits that can be obtained by the customer. More recently, Kohlborn, et al. (2009) have defined a service as "an autonomous transformational capability that is offered to and consumed by external or internal customers for their benefit".

In the last years, also the concept of Service System became relevant, due to the complexity of the environment and the high number of the interaction characterising services (Visintin, et al., 2014). For example, Katzan (2011) posited that "a service system is a socially constructed collection of service events in which participants exchange beneficial actions through a knowledge-based strategy that captures value from a provider-client relationship".

These statements describe how services have been traditionally perceived and defined as something intangible, and their nature viewed as an activity or process (Johns, 1999). Therefore, it is possible to adopt (or partially adapt) commonly used modelling approaches, paradigms and tools (being simulation one of them) to (re-)design and analyse services and their performance.

6.2.2 Service engineering

Considering the wide spread of services and the different sectors addressed, their nature and diversity are such that a systematic design approach is required. In this sense, the term Service Engineering (SE) was coined in the mid-nineties to refer to a technical discipline concerned with the systematic development and design of services using suitable models, methods and tools (Bullinger, et al., 2003; Shimomura & Tomiyama, 2005), aiming at filling the gap in the evolution between product and service design. It mainly aims to improve the service planning, service conception and service implementation planning, in order to create more valuable services (Pezzotta, et al., 2012).

However, many definitional elements, methods and tools are still under development (Cavalieri & Pezzotta, 2012) and relatively few methods are available (e.g. service blueprinting (Bitner, et al., 2008)). Furthermore, the majority of models, methods and tools adopted for service design derive from product and software engineering (e.g. TRIZ, QFD, FMEA (Baines, et al., 2007; Alonso-Rasgado, et al., 2004; Van Halen, et al., 2005; Kett, et al., 2008; Rapaccini, et al., 2013; Aurich, et al., 2006; Shimomura & Tomiyama, 2005)). Services by definition exist in the moment in which they are delivered to the customers through a process (intangibility). The process is realized together by the provider and the customer (inseparability) and cannot be kept in stock (perishability) (Bartolomeo, et al., 2003; Fitzsimmons & Fitzsimmons, 2000; Grönroos, 1990). Thus, these features make the existing product and software engineering approaches unsuitable to design properly services (Cavalieri & Pezzotta, 2012).

For these reasons, some authors in the SE research domain tried to develop specific methods – or to tailor and adapt the existing ones – in order to better consider the service features. Most of the developed methods and tools focus on the satisfaction of customers' needs, as required by the service definition. Customer satisfaction itself, however, cannot grant long-term company sustainability and profitability. In order to make service provision profitable in the long term, it is of utmost relevance to balance the excellence in the service provision with a high efficiency and productivity of the service processes. Indeed, one of the most relevant issues in the context of SE is the identification of a proper balance between customer value and company efficiency (Cavalieri & Pezzotta, 2012).

In order to gather the dynamism and balance the customer and the company perspective, business process simulation can be adopted to support engineering and early assessment of a service. In particular, given the definition of a "service as a process", discrete event simulation (DES) paradigm, traditionally used for the analysis of manufacturing processes, can still represent a valid choice for service process analysis. Thus, DES can be used as a design tool both during the service engineering process, since it provides a support for qualitative and quantitative assessment of company decisions, and during the service re-engineering process, since it can be used as a decision making tool for the improvement of the actual service delivery processes (Visintin, et al., 2014;

Chalal, et al., 2015; Babulak & Wang, 2007; Rondini, et al., 2015; Pezzotta, et al., 2014, 2015).

In the remainder of this chapter, we focus on the simulation of a pure service delivery process for SE purposes, providing a comparative analysis of available simulation software solutions and paradigms suitable for service delivery process simulation.

6.3 SIMULATION IN THE SERVICE CONTEXT

Simulation is defined as "the process of designing a model of a real system and conducting experiments on this model for understanding the behavior of the system and/or evaluating various strategies for the operation of the system" (Shannon, 1998; Banks, 1998). Traditionally, simulation has been mainly focused on production processes in manufacturing, supply chain management and service operations due to its potential benefits. However it has not been applied extensively to support the definition of new service during the design and engineering phases. Simulation can provide insights into an existing or a hypothetical situation, allowing for a safe, replicable, and usually less expensive performance test and analysis compared to a real-scale implementation, where in the majority of the cases a trial and error approach is not a viable option (Hlupic & Robinson, 1998; Laughery, et al., 1998). These benefits of simulation have contributed to achieve significant improvements in productivity and quality of manufacturing processes.

Recently, simulation started to play a relevant role in the engineering and the design phases of a service due to the increasing competitive pressure to reduce the time to serve customers, to obtain higher profits, and develop new services. Simulation approaches can be used for the analysis of any system to ensure quality, timeliness, and efficiency of stochastic, complex processes that operate in resource-constrained environments. In addition, it boasts the following advantages (Hlupic & Robinson, 1998):

- Flexibility. Simulation allows the investigation of many different situations, regardless how complex they are.
- Ease of understanding. The concept on which simulation is based is simple, referring to the replication of a real situation through a suitable model.
- Provide insights. In a model, it is possible to observe the delays occurred during the simulation and identify bottlenecks, for example in information, material and product flows.
- Answer questions. Simulation helps to understand how or why certain phenomena occur in the analysed system.
- Quick identification of relevant variables. In each simulation model, it is possible to define the most important variables to control and to increase performance.
- Manage time. Having control over time and monitor the situation during several years is very useful when observing performance in the long term.

Given the definitions of service, along with the fact that most services are fairly well defined discrete processes, DES offers great potential as a means of describing, analyzing, and optimizing service and service systems of many types (Laughery, et al., 1998) and supports their systematic and optimized engineer.

However, the KPIs to analyse and the variables to monitor in a service delivery process are quite different from the traditional manufacturing process; indeed the concept of a good "service" includes the consideration of the quality and timeliness of service performance. In addition, what is intended with service productivity is not clear yet (Gronroos & Ojasalo, 2002). However, the tight relationship between the customer and the provider requires the use of novel methods, which go beyond a static description of work processes. In particular, the representation of the underlying dynamics and uncertainties of a work organization should enable service managers to simply generate valid service scenarios (Duckwitz, et al., 2011). Therefore, simulation can allow to identify and measure the dynamic performance of a given process, and to assess both the presence and relevance of any queue and/or bottlenecks. Finally, simulation can also allow for the better understanding of the causes of the dynamics that emerge during the execution of the service process.

6.3.1   Main aspects of simulation in the service domain

As previously underlined, service delivery processes have some features that differentiate them from a traditional manufacturing process. In this paragraph, the description of a service process through simulation is presented, and the specific services' features with respect to the simulation are highlighted.

All service delivery processes and related support processes can be considered as business processes. A generic business process is a set of functions or tasks, logically linked together, using the organization's resources to achieve the ultimate goals. In the service, entities that execute, or participate into, these processes (customer requests, orders, etc.) revolve around a series of service stations (service delivery activities). Once the entities have completed all the activities in the process, the service is delivered and the entities disposed. Although to a certain degree, this working mechanism is similar to manufacturing systems, service delivery processes have some unique characteristics making the simulation model more complex. The following differences between simulation in manufacturing and in the service domain can be identified (Banks, 1998):

–   Arrivals in a service are typically random and cyclical: the arrival of requests for a service can be correlated with the day or the time within a day, for example. The inter-arrival times (i.e. the time between two consecutive arrivals) can be usually well represented by an exponential distribution (Gladwin & Tumay, 1994).
–   Resources are prevalently people: differently from machines, resources represented by people have different performance and characteristics. For example, the productivity of a worker changes during the working shift; people are subject to interruptions, and have less predictable behavior when confronted with unforeseen events. According to (Duckwitz, et al., 2011), human beings have three features that are difficult to simulate: i) decision making behavior (the decision-making process can be influenced by deadlines and rewards, and therefore not being completely rational); ii) cooperative work (different humans have different attitudes toward collaboration); iii) human reliability (related to personal, organizational and environmental factors). In addition, human resources may have different skills and qualification, and in many cases they are moving resources (i.e. they are required to move to customer premises in order to deliver the service, as

in on-site technical assistance services) (Lagemann, et al., 2015). Therefore, due to the prevalent presence of people, system variability in service is often much higher than in systems with automated resources and, therefore, of greater importance.

- Entities are often people: similar to resources, entities being served often represent people that may have a rather complex behavior and preferences that are difficult to predict. These behaviors change dynamically during the execution time of an activity and waiting times, as well as the saturation of resources (Lee, et al., 2015). Considering a queuing system, a customer can enter the system but leaving before joining a queue, for example because it is too long (balking). Otherwise, a customer can join a queue and leave before being serviced because he/she has waited for too long (reneging). Finally, a customer can join a queue and then move to another, seemingly faster one (jockeying) (Chung, 2004). Due to such a possibility in customer choice, system variability can also increase significantly.
- Process times are highly variable: especially when resource and entities are people, the processing time may depend upon the status of the system. In a service delivery process, the value is created from the interaction between the resources (representing the service provider) and the entities (customers). The uncertainty related to those people increases exponentially while allowing their interaction. For example, an increment in a queue length can push the resources to work faster, or a change in the state of an entity (the arrival of an angry customer) can push resources to work faster to complete the service (Gladwin & Tumay, 1994). All these factors are hard to represent and predict with a simulation model.
- Lack of steady-state behavior over the whole running period: due to the random and cyclic arrivals, service usually does not reach steady states. Moreover, arrivals may change according to the day or the time within the day. Due to this, it would be necessary to simulate the system in each period separately.
- There is often no clearly defined set of components as in manufacturing: the simulation package must often define system behavior without the use of hard data on the process. For example, in modelling a manufacturing process, the analyst will usually have access to drawings, specifications and production organization. Service environment will rarely have such refined documentation. The simulation modeler is often called upon to define and understand the process in a more complete way than anyone did before.
- Waiting time tends to have a much greater importance than throughput: a service cannot be stocked. Therefore, the process and waiting times have great relevance in a service environment: people hate to wait to be served. In many services, time in the queues will always be the key measure of performance and excessively long waits will not be tolerated (Abbasi, et al., 2013).
- Services are often short-term demand driven, and these demands can vary by day and time: fluctuations in demand (in terms of both mix and quantity) for services will greatly affect the service's ability to provide good service. Again, because of the variability not only of service providers but also of those demanding services, it is important to understand and predict the factors affecting customer demand.

Besides these notable differences, services are no distinct from any other type of system. To some extent, a process is a process, and the flow of entities through a process with constrained resources is fundamentally the same for services and manufacturing systems. At a certain level of abstraction, there is no difference between a lathe and a bank teller: both have to perform a set of action to serve an entity. Furthermore, at the root of virtually all types of simulation study, the question remains the same: how can we do more with less? Although the particular issues and emphases may differ, the basic issues are the same with manufacturing, service, or any other type of discrete system.

6.3.2 Existing work related to simulation and services

Due to the reasons highlighted before, the existing body of work related to the simulation of service delivery processes is poorer than in the manufacturing domain (Babulak & Wang, 2007; Visintin, et al., 2014; Pezzotta, et al., 2015). Hereafter, a brief summary of existing research in the area is presented.

Discrete Event Simulation (DES) has been widely used in modelling health-care systems for many years (Günal & Pidd, 2010) and different software have been adopted. However its adoption in other service sectors is still limited. Watanabe et al. (2012) adopted a simulation approach (Tateyama, et al., 2009) to evaluate a service process and verify the described approach with a test case in a bike rental service.

Harpring et al. (2014) adopted DES to simulate a social service. In particular, they adopted the simulation to investigate various alternatives concerning process flow, staffing, layout, and to carry out experiments about possible changes. Pezzotta et al. (2014, 2015) and Rondini et al. (2015) adopted DES as a foundation element of the Service Engineering Methodology (SEEM) to evaluate the service component of a PSS and to build a what-if analysis to define alternative processes.

Visintin et al. (2014) proposed a simulation study in the aerospace industry concerned with the design of a service delivery system. Hirth et al. (2015) tried to define possible applications of dynamic models to the service component of a PSS. They used DES to allocate the proper amount of resources in a car sharing system. Yoon et al. (2012) defined an evaluation method for a PSS solution. They used DES to evaluate the competition of a company in the market.

Furthermore, additional works related to the adoption of simulation can be found in literature, but they refer to others simulation paradigms, such as agent-based modelling and system dynamics. Due to our focus on DES, we excluded those paradigms from our current analysis.

6.3.3 Simulation software for the service field

Given the availability of different DES software packages, we focused on the comparison of four commonly used packages with the aim to understand their strengths, weaknesses and overall suitability with regard to the SE domain. The four have been selected in order to compare software with distinguishing features and characteristics. Therefore, the analysis would show what the most suitable for Service Engineering purpose. The software compared are the following (Table 6.1):

–   Software A that has been used by the majority of the application found in literature and is one of the most adopted in the service industry (Abu-Taieh & El Sheikh, 2007);

Table 6.1 Main characteristics of the simulation software compared.

Software package	Main characteristics
Software A	– Graphical modelling/animation system – Based on hierarchical modelling concepts – Power and flexibility of the simulation language
Software B	– Modelling and simulation of production and logistics systems and processes – Provides bottlenecks analysis, statistics and charts
Software C	– Add on of a traditional modeling tool supporting BPMN – Simulate static flowchart and workflow diagrams – Support "what-ifs" analysis natively – Graphical representation of outputs
Software D	– Modelling method for discrete-continuous hybrid systems – Graphical User Interface supporting the designers in to edit and simulate the models developed – Still under development

- Software B, another well-adopted software whose features are specifically developed for the manufacturing sector;
- Software C, a quite recent simulation software that is an add-on of a traditional modelling tool that supports the Business Process Modeling Nomenclature, which is the most-used nomenclature in modeling services;
- Software D, to our current knowledge the only software specifically developed for SE purposes.

6.4 THE COMPARISON METHODOLOGY

In order to compare the four software packages listed in Table 6.1, we defined a common test case; in particular, we referred to a real-scale case located in a take-away food shop offering home-delivery service (see section 4.2 for a description of the case). Such a test case represented the base situations on which the four software packages have been compared to understand their main features and their practical applicability. To this end, for each simulation software package, the following activities were performed:

- Development of the service process model: using the service blueprinting, a technique for service model visualization (Bitner, et al., 2008) and the Business Process Modeling Notation (BPMN), a process flow diagram was developed. Due to the specificity of the service blueprinting, the process model puts into evidence the different actors involved, the relations among actors as well as the customer processes.
- Data and distribution identification and integration into the model: to obtain a working model, quantitative data and distribution information of the process (i.e. task duration times, process alternatives frequency, resource usages) were

integrated into the model according to the software functionalities provided by each software package (i.e some software provide a data fitting tool for distribution identification, whereas for the other software the input has been calculated separately). For the sake of accuracy, the service blueprinting structure, which divides the customer, the front office and the back office actions, was kept as much as possible.

–   Model validation: once all the data were included in the model and the simulation run, the results provided were compared with the real data to validate the model. Consequently, to this activity, the model was amended to reflect the real case with reasonable accuracy.

6.4.1　Evaluation criteria for the simulation packages

To the authors' best knowledge, besides software D, there are no commercial software packages specifically developed for the simulation of complex services that can effectively support the design and development of service-delivery processes by using concepts from the SE domain. Therefore, the comparison of the four software packages was aimed to understand which is the most suitable in simulating a pure service process. To this end, we defined a list of criteria gathered from the literature.

In the last, recent years, many contributions have stated their preferred list of important criteria for simulation software selection (Banks, 1991; Banks, et al., 1991; Davis & Williams, 1994; Hlupic, 1997; Bosilj-Vuksic, et al., 2007; Abu-Taieh & El Sheikh, 2007). However, it seems that there is still a lack of a standard list. Due to its widespread adoption and the scope of the research, Nikoukaran & Paul's (1998) list of criteria and sub-criteria, summarized in Table 6.2, was adopted in this study.

6.4.2　The food shop case

The analysed case refers to a food shop (more specifically, a *pizzeria*) that offer both take-away food and home delivery services. The shop process works as follows.

The customer can either call the shop on the phone or show up directly in the pizzeria to place his/her order. Once the customer has placed the order, the clerk assesses and negotiates with him/her the service (waiting) time. If the negotiation terminates successfully (i.e. the customer is willing to wait for the order), the order is allocated according to schedules and priorities, and then it is transmitted to the pizza-maker. Every day, 150 pieces of pizza dough are available. Once assured that there are enough pieces of dough, the pizza-maker proceeds to fulfil the orders. If not enough dough is available the order is refused and the clerk calls back the customer. In case some ingredients are depleted before the end of the service, the clerk has to prepare and replenish them.

The next step is to cook the pizzas. If done correctly, the pizzas are put into the boxes. Otherwise, the order returns at the beginning of the preparation stages. If the customer is in the shop, the clerk delivers the pizzas in the boxes to the customer, and receives the payment in return. Otherwise, the delivery boy delivers the boxes directly to the customer's house.

Table 6.2 Evaluation criteria for the simulation software packages.

Criteria	Definition	Sub-criteria
Vendor	This criterion is for evaluation of the credibility of the vendor, and to some extent his/her software	Pedigree Documentation Support Pre-purchase
Model and Input	This category of criteria includes issues related to a model, its development and data input	Coding aspects Queuing Policies Statistical distributions Input Library of reusable modules Model building
Execution	This criteria group includes issues related to experimentation	Multiple runs Automatic batch: Reset capability Start in non-empty Speed control
Animation	Criteria for evaluation of animation deal with creation, running and quality of animation	Icons Screen layout Development Running
Testing and Efficiency	This category can be used to evaluate testability, debugging power and efficiency of a package	Validation and verification Multitasking Interaction Step function Breakpoints Backward clock Conceptual model generator Limitations Display feature Tracing
Outputs	This criterion covers the issues related to how the output are shown	Reports Delivery Integration Data base Graphics Analysis
User	The user criteria group deals with some specific user needs and circumstances	Simulation type Orientation Hardware Operating system: Network version Security device Required experience Financial: Software class

(Source: Nikoukaran & Paul, 1998)

Thus, the case involves four actors: the customer, the clerk, the pizza-maker, and the delivery boy. The process structure and the data have been collected thanks to on-site surveys and measurements.

6.5   MAIN RESULTS AND DISCUSSION

The described case was modelled and simulated in the four software packages (Table 6.1). The comparison of the packages is based on the evaluation criteria proposed in Nikoukaran & Paul (1998) and considering the service specific characteristics, as previously described. Hereafter, the main outcomes are reported and discussed.

- Unpredictable and cyclic nature of the arrivals: all the simulation packages are able to manage this aspect. However, software A and C seem more suitable because of their flexibility and better management of the day-by-day scheduling through calendars. The definition of day-by-day schedule was much easier with Software C's interface but, on the other hand, software A has the "input analyzer" tool that facilitates the definition of arrival distribution given a set of historical data.
- Resources are prevalently people: resources are managed in different ways in the different packages. With software A and C, a resource should be assigned to each activity, and released at the end of the task. In software A, there is also the possibility to group different kinds of resources that can perform the same task (parallel working) while in Software C this feature is more complex to implement. Concerning Software B, it works considering the resources skills: a skill is assigned to the worker and then the process uses the resources with the appropriate skills. Software D does not differentiate between entities and resources; therefore, to make the system working as requested, several additional conditions have to be set in the model.
- Variability of demand: the variability of the process time was considered and simulated by a proper use of the scheduling feature. Most of the consideration and statistical analysis were done in order to depict the behavior of the customers and resources in the process. As mentioned before, software A and C provide a better tool for managing the scheduling.
- Process times are highly variable: waiting time and the related queues have a great importance in a pure service process. Regarding this aspect, software A, thanks to the "Queue" module, has an effective and easy queue management. Software C can also tackle this aspect upon quick definition of the user. Software B can also handle this issue if properly modelled. On the other side, Software D does not consider this aspect.

For what concerns the list of criteria previously reported and identified in literature, we can state the following:

- Model and Input: by comparing the development process of the models, Software C is preferable because it allows the direct simulation of the BPMN blueprinting model. This also prizes the visual impact of the model since there are no graphical restrictions. In software A, the visual impact and the software interface allow reproducing a service process model using its basic modules. These processes reflect and remind effectively the categories' elements identified in the flow model description done with the service blueprint of the BPMN. In software

B, function related functions mainly remind machine tools or a workstation of a manufacturing process. For example, the decision-making activity *"account pizza"* of the pizza-maker is more abstract than physical; however, to simulate it with Software B, it is necessary to use a real workstation. Software D has fewer modules to edit the model (it has just actors, scenes, transactions and strings); this entails more time and a deeper knowledge of the tool for an effective model development.

– Execution: the management of the simulation (start, stop, speed, warm-up period) are perfectly handled by software A, B and C. However, the event control tool of software B is easier and more intuitive than the setup run of the other two commercial software. Setting the simulation speed in software D is very difficult because it is impossible to do it from the toolbar, and it is necessary to stop the run manually: therefore, a "stop time" of the simulation cannot be set precisely.

– Animation: software A allows changing easily the icon of the entity that goes through a part of the model, facilitating the understanding of the flow. For example, if the *"order"* entity moves from the clerk to the pizza-maker, it is possible to change the image from a post-it to a pizza (there are some types of graphic representation in the library, but freehand shapes of low resolution can be drawn). Furthermore, it is easy to recognize the entities queuing in the process. Software B and C have a good animation of resources, because they allow understanding where the worker is located. Software D presents basic animation, it is therefore difficult to follow the model during the simulation: it is impossible to change the zoom of the window; therefore, for large models, it is impossible to have a clear view. Moreover, the tokens movement creates confusion between entities and resources.

– Output: simulation packages can develop excellent statistical outputs. All the three commercial software allows customization of the output statistics. All of them, however, require extra-knowledge to get the data. For example, in Software B, to see the workers saturation, it is necessary to use external libraries, load an appropriate tool and elaborate summary graphs. In addition, it requires the creation of specific tables with the specific language. There is also the "report" tool that creates printable reports; however, the outputs obtained are generic and not specific to study a service process. Software A and C automatically generate reports at the end of the simulation, which are more appropriated for the analysis of a pure service. Queuing, process and waiting times are the main outputs. They can both allow the collection of additional and customized output. Software C also shows some graphs. Finally, one of the main advantages of Software C is the easy export of the results in Excel that directly allows analysis of data. The export format of software A in excel is more difficult to use and requires additional formatting adjustments. In Software D, the outputs are not printed automatically: it is possible to obtain a graphical representation of the different analyzed variables and attributes (value and time). By setting an appropriate number of global variables and taking advantage of the differential equations (accepted by the software), it is possible to calculate the use of resources and the waiting times of the entities. However, it is not easy, especially because it currently allows the creation of only ten global variables.

–   User: it is easier to find documentation and examples in software A related to the simulation of the service process. Even though it is more difficult to learn and properly manage, Software B, thanks to the programming language, has the possibility to integrate pieces of code manually. This makes the simulation package more flexible. Furthermore, Software B allows to control the decide module, namely the management of conditional flows through the object-oriented programming, record any type of values (statistical, inter-arrival, availability) into exportable tables, and interact with the simulation through Buttons, Prompt, Communications, etc. In software A, instead, the specific code is automatically generated when the modules are placed in the diagram window. It is possible to insert some controls (through Visual Basic Macro) but it is more laborious than the solution offered by software B with the internal language. Similar to software A, software C automatically defines some features of the modules, but additional specifications can be provided with a more complex interface that requires some programming skills. Software D presents several glitches, and usage issues attributable to its early stage of development. Nonetheless, it allows using simulation without a deep knowledge of the subject. Furthermore, its orientation to service engineering is promising for the next releases.

Finally, considering modelling flexibility and easiness, software A, B and C all provide a great flexibility in modelling. However, in software A and C the model development was easier. Software A package can be easily adapted to different types of scenario because of the wide choice of simulation modules. Software C, being an add-on of a traditional modelling tool, allowed a varied range of possible choices for simulating an activity. A "service" module is also available. On the other hand, modelling a service environment in software B resulted a complex task, due to its prevalent product-orientation focus. Indeed, this package gives emphasis to statistics and failure management, set-up time, MTTR, etc., which are generally not relevant in simulating a service. Nonetheless, if proper proxies are stated, and the simulation is properly managed, it is possible to reproduce effectively and efficiently a complex service process. The development in the software D environment was complex, due to the considerable object-oriented skills required, and to the limitations of the actual experimental version of the software.

In conclusion, it is possible to highlight the advantages of commercial software with respect to the software D that is still under development. Software B is strongly oriented to manufacturing, and this implies some disadvantages related to model definition and output. Therefore, software A and C are the two most appropriate software packages since they provide general output, and they are easier to deal with. However, it is worth recalling that the main advantage, that can differentiate the evaluation of the four software packages, is the integration of software C with a modelling tool that supports BPMN nomenclature. This allows the development of the Service Blueprinting by use of BPMN. This is a strong advantage in terms of clarity and visualization of the model. Concerning other parameters, the software C can be compared to software A in terms of both clarity and ease of use. In the Table 6.3 hereafter, a brief summary the most relevant and distinguish criteria of the evaluation is reported.

Table 6.3 Summary of the simulation packages evaluation.

	Criteria	Software A	Software B	Software C	Software D
Specific service requirements	Randomness and cyclic of the arrivals	++	+	++	−
	Resources are often people	+	++	+	=
	Variability of demand	+	=	+	−
	Process times are highly variable	++	=	++	=
Simulation package features	Model and Input	+	−	++	=
	Execution	+	+	+	=
	Animation	++	+	+	−
	Output (Reports)	+	−	++	=
	Output (Integration)	−	+	+	=
	User	+	+	+	+

(Legend: ++ marked superiority; + moderate superiority; = aligned with the expectation; − moderate inferiority; −− marked inferiority)

6.6 CONCLUSIONS

In the SE context, there is the need to develop tools supporting the design and the assessment of the service delivery processes. This work tries to support research in this direction while performing a comparative analysis among four different simulation software (A, B, C and D) in a pure service context. To understand better how the four selected simulation packages could be adopted in a service environment, they were adopted to model a take-away food shop. Two different sets of comparison criteria have been evaluated: criteria evaluating specifically how the software can be used in a service environment and the evaluation criteria proposed by Nikoukaran and Paul (1998) which generally support the comparison of simulation software.

Considering the results of our analysis, Software C can be considered the most suitable software to simulate a service delivery process, mainly because of its interface with the modelling tool that allows the creation of a service blueprinting map representing all the service features that, in turn, can be run as a simulation model. Concerning the ease of implementation and the capability to elaborate data with a service-oriented perspective, software C has similar characteristics to software A. However, independently from the simulation package selected, the simulation of service delivery processes implies some critical points: in particular, during our analysis, some parameters were difficult to set and collect.

For example, services often require the interaction with customers, who may introduce delays and increase processes variability. Furthermore, human decision-making processes, unless based on a specified set of rules strictly followed by the operators, are often too complex and convoluted to be modelled and simulated. Therefore, in our case customer's behavior and decision-making processes have been approximated via a random decision process, imitating the historical behavior of the customers and operators.

For these reasons, this analysis cannot be deemed as complete and exhaustive. The current work could be enlarged and completed with the identification of other possible simulation packages. In addition, evidence from existing literature in the field shows that DES can be integrated or substitute by other simulation paradigms. Therefore, a comparison related to the analysis of other existing simulation paradigms together with the associated software could complete the current work.

REFERENCES

Abbasi, H., Rahimi, F. & Senobari, M., 2013. The Impact of Waiting Time on Customer Satisfaction & Loyalty in the State and Private Banks in Tehran. *World of Sciences Journal,* pp. 1(16): 161–174.

Abu-Taieh, E. & El Sheikh, A., 2007. Commercial simulation packages: a comparative study. *International Journal of Simulation,* pp. 8(2): 66–76.

Alonso-Rasgado, T., Thompson, G. & Elfström, B., 2004. The design of functional (total care) products. *Journal of Engineering Design,* 15(6), pp. 515–540.

Aurich, J., Fuchs, C. & Wagenknecht, C., 2006. *Modular design of technical product–service systems.* The Netherlands, Brissaud D., et al., pp. 303–320.

Babulak, E. & Wang, M., 2007. Discrete event simulation: State of the art. *International Journal of Online Engineering,* 4(2), p. 60.

Baines, T. et al., 2007. State-of-the-art in product-service systems, Pro-ceedings of the Institution of Mechanical Engineers. 221(10), pp. 1543–1552.

Banks, J., 1991. *Selecting simulation software.* Arizona, USA, s.n., pp. 15–20.

Banks, J., 1998. *Handbook of Simulation: Principles, Methodology, Advances, Applications, and Practice.* Canada: Wiley – Interscience.

Banks, J., Aviles, E., McLaughlin, J. & Yuan, R., 1991. The simulator: new member of the simulation family. *Interfaces,* pp. 21(2), 76–86.

Bartolomeo, M. et al., 2003. Ecoefficient producer services e what are they, how do they benefit customers and the environment and how likely are they to develop and be extensively utilised? *Journal of Cleaner Production,* pp. 11, 829–37.

Berry, L. L., 1980. Service marketing is different. pp. May–June, 24–29.

Bitner, M. J., Ostrom, A. & Morgani, F., 2008. Service Blueprinting: a practical technique for service innovation. *California Management Review,* p. 50(3): 66–94.

Bosilj-Vuksic, V., Ceric, V. & Hlupic, V., 2007. Criteria for the evaluation of business process simulation tools. *Interdisciplinary Journal of Information, Knowledge, and Management,* pp. 2, 73–88.

Bullinger, H.-J., Fähnrich, K.-P. & Meiren, T., 2003. Service engineering-methodical develop-ment of new service products. *International Journal production Economics,* pp. 275–287.

Cavalieri, S. & Pezzotta, G., 2012. Product-Serice Systems Engineering: State of the art and research challenges. *Computers in industry,* pp. 278–288.

Chalal, M., Boucher, X. & Marques, G., 2015. Decision support system for servitization of industrial SMEs: a modelling and simulation approach, J. *Journal of Decision Systems.*

Chesbrough & Spohrer, 2006. A research manifesto for service science. *Communications of the ACM,* Vol. 49, No. 7.

Chung, A., 2004. *Simulation Modelling Handbook. A practical approach.* s.l.: CRC Press LLC.

Cook, D., Goh, C. & Chung, C., 1999. Service typologies: a state of the art survey. *Production and Operation Management,* pp. 8(3), 318–338.

Davis, L. & Williams, G., 1994. Evaluation and selecting simulation software using the analytic hierarchy process. *Integrated Manufacturing Systems*, pp. 5(1), 23.

Duckwitz, S., Tackenberg, S. & Schlick, C., 2011. *Simulation of human behavior in knowledge-intensive services*, RWTH Aachen University, Germany: Institute of Industrial Engineering and Ergonomics.

Edwardsson, Gustafsson & Roos, 2005. Service Portraits in Service Research: A critical Review. *International Journal of Service Industry Management*, pp. 107–121.

Fitzsimmons, J. & Fitzsimmons, M., 2000. *New Service Development: Creating Memorable Experiences*, s.l.: Thousand Oaks, CA: SAGE Publications.

Gladwin, B. & Tumay, K., 1994. *Modeling Business Processes with Simulation Tools.* s.l., J.D. Tew, S. Manivannan, D.A. Sadowski, and A.F. Seila.

Goldstein, S. M., Johnston, R., Duffy, J. & Rao, J., 2002. The service concept: the missing link in service design research?. *Journal of Operations management*, pp. 121–134.

Grönroos, C., 1990. *Service Management and Marketing: Managing the moments of truth inservice competition*, s.l.: Lexington: Lexington Books, p. 27.

Gronroos, C., 2000. *Services Management and Marketing: A Customer Relationship Approach*, UK: Wiley.

Gronroos, C. & Ojasalo, K., 2002. Service productivity: towards a conceptualization of the transformation of inputs into economic results in services. *Journal of Business Research*, pp. 57(4), pp. 414–423.

Günal, M. M. & Pidd, M., 2010. Discrete event simulation for performance modelling in health care: a review of the literature.. *Journal of Simulation*, 4(1), pp. 42–51.

Harpring, R., Evans, G. W., Barber, R. & Deck, S. M., 2014. Improving efficiency in social services with discrete events simulation. *Computers and industrial engineering*, pp. 70, pp. 159–167.

Hirth, N. et al., 2015. *An approach to reveal starting point for PSS design support with dynamic models.* s.l., Proceedings of the 7th Industrial Product-Service System Conference-PSS, industry transformation for sustainability and business.

Hlupic & Robinson, 1998. *Business Process Modelling and Analysis Using Discrete-Event Simulation.* s.l., s.n.

Hlupic, V., 1997. Simulation software selection using SimSelect. *Simulation*, pp. 69(4), 231–239.

Johns, N., 1999. What is this thing called service? *European Journal of Marketing*, pp. 33(9/10), 958–974.

Katzan, H., 2011. Foundations of service science concepts and facilities. *Journal of Service Science*, 1(1), pp. 1–22.

Kett, H., Voigt, K., Scheithauer, G. & Cardoso, J., 2008. *Service engineering in business ecosystems.* Stuttgart, Germany, Fraunhofer IRB Verlag.

Kohlborn, Fielt, Korthaus & Rosemann, 2009. *Towards a Service Portfolio Management Framework.* Melbourne, s.n.

Lagemann, H., Boblau, M. M. & Meier, H., 2015. *The influence of dynamic bsuiness model on IPS2 network planning – an agent-based simulation approach.* Saint Etienne, s.n.

Laughery, K. R. et al., 1998. Effects of Warnings on Responsibility Allocation. Psychology & Marketing. pp. 15, 687–706.

Lee, S., Han, W. & Park, Y., 2015. Measuring the functional dynamics of product-service system: A system dynamics approach. *Computers & Industrial Engineering*, pp. (80) 159–170.

Marc, S., 2012. *This is service design thinking: Basics-Tools-Cases.* s.l.: Bis Publishers.

Nikoukaran, J. & Paul, R., 1998. Simulation software selection "whys and hows". *Yugoslav Journal of Operations Research*, pp. 8(1), 93–102.

Parker, D. W., 2012. *Service operations Management. The total experience.* Cheltenham (UK): Edward Elgar Publishing Limited.

Pezzotta, G., Cavalieri, S. & Gaiardelli, P., 2012. A spiral process model to engineer a product service system: an explorative analysis through case studies. *CIRP Journal of Manufacturing Science and Technoloy*, pp. 5(3): 214–225.

Pezzotta, G., Pinto, R., Pirola, F. & Ouertani, M. Z., 2014. Balancing Product-service provider's performance and customer value: the Service Engineering Methodology (SEEM). *Procedia CIRP*, Volume 16, pp. 50–55.

Pezzotta, G. et al., 2015. A Service Engineering framework to design and assess an integrated product-service. *Mechatronics*.

Ponsignon, F., Smart, P. A. & Maull, R. S., 2012. Process design principles in service firms: Universal or context dependent? A literature review and new research directions. *Total Quality Management & Business Excellence*, 23(11–12), pp. 1273–1296.

Quinn, J., Baruch, J. & Paquette, P., 1987. Technology in services. *Scientific American*, pp. 257(6), 50–58.

Rapaccini, M. et al., 2013. Service development in product-service systems: a maturity model. *Serv Ind J*, pp. 33(3–4): 300–319.

Rondini, A. et al., 2015. *Service Engineering Methodology in Practice: A case study from power and automation technologies*. Saint Etienne, France, s.n.

Rondini, A. et al., 2015. Business Process Simulation for the Design of Sustainable Product Service Systems (PSS). *Advances in Production Management Systems: Innovative Production Management Towards Sustainable Growth*.

Shannon, 1998. *Introduction to the Art and Science of Simulation*. s.l., s.n.

Shimomura, Y. & Tomiyama, T., 2005. Service modeling for service engineering. In: *Knowledge and Skill Chains in Engineering and Manufacturing*. s.l.: s.n., pp. 31–38.

Tateyama, T., Shimomura, Y. & Kawata, S., 2009. *Development of Scene Transition Nets (STN) GUI Simulator for Service Flow Simulation*. Palo Alto (CA), s.n., pp. 337–346.

Teixeira, J. et al., 2012. Customer experience modeling: from customer experience to service design. *Journal of Service Management*, 23(3), pp. 362–376.

UK Design Council, 2010. *What is service design?* [Online] Available at: http://www.designcouncil.org.uk/about-design/Types-of-design/Service-design/What-is-service-design/

Van Halen, C., Vezzoli, C. & Wimmer, R., 2005. *Methodology for Product Service System Innovation How to implement clean, clever and competitive strategies in European industries.*, Assen, Netherlands: Royal Van Gorcum.

Vandermerwe, S. & Rada, J., 1988. Servitization of Business: Adding Value by Adding Services. *European management Journal*.

Visintin, F., Porcelli, I. & Ghini, A., 2014. Applying discrete event simulation to the design of a service delivery system in the aerospace industry: a case study. *Journal of Intelligent Manufacturing*, 25(5), pp. 1135–1152.

Watanabe, K., Mikoshiba, S., Tateyama, T. & Shimomura, Y., 2012. Service process simulation for integrated service evaluation. *Journal of Intelligent Manufacturing*, 23(4), pp. 1379–1388.

Xiaofei, X. & Wang, Z., 2011. State of the art: business service and its impacts on manufacturing. *Journal of Intelligent Manufacturing*, 22(5), pp. 653–662.

Yoon, B., Kim, S. & Rhee, J., 2012. An evaluation method for designing a new product-service system. *Expert system with application*, pp. 39, 3100–3108.

Zeithaml, V. A. & Bitner, M. J., 1996. Services Marketing. *New York: McGraw-Hill*.

Chapter 7

The impact of power structure on service supply chain management

Xiaojun Wang[1] *& Xu Chen*[2]

[1] *School of Economics, Finance and Management, University of Bristol, Bristol, UK*
[2] *School of Management and Economics, University of Electronic Science and Technology of China, Chengdu, P. R. China*

SUMMARY

Service supply chain management has attracted increasing attention from academics and practitioners because of its importance to the world economy. Operational decisions made by service firms are often influenced by the power relationship in a service supply chain. This research investigates the impact of different power structures on firms' operational decisions and performances in the context of two different service supply chains. Two case studies are presented in the chapter. The first case focuses on a retail service supply chain with an offline and online mixed dual-channel. The second case analyses a mobile phone service supply chain with a free and bundled dual-channel. We employ the game theoretical approach to analytically examine the impact of different power structures on the pricing and channel selection decisions as well as the firms' and the supply chains' performances in both case studies. Our findings provide many interesting insights, which will be beneficial for firms to make important strategic and operational decisions and improve their performances.

Keywords

service supply chain management, mobile phone, retail operations, dual-channel, pricing, game theoretical approach

7.1 INTRODUCTION

Service supply chain management has received an increasing attention by both academics and practitioners because of the importance of the service sector to most economies in the world (Spring and Araujo 2009; Stavrulaki and Davis 2014; Wang et al. 2015). There are different types of service supply chains depending on the relative goods/services composition of products. In addition to service only supply chains, many more supply chains such as restaurant and food retail supply chains manage physical products together with significant service considerations (Wang et al. 2015). Service supply chains impose some unique challenges as compared to conventional manufacturing supply chains. For instance, customer need is often influenced by the price and quality of services as well as the price and quality of physical goods if they are also a part of product package. On one hand, it requires cooperation between service provider and goods producer to meet end consumers' needs. On the other hand, there is a competitive relationship between them when distributing the profit gained from end consumers.

Adding to the complexity of the problem, operational decisions such as ordering, quality, inventory, and pricing made by service providers and their suppliers are often influenced by their relative power in the supply chain against their customers, suppliers or market competitors. A dominant power in the supply chain does not only enable companies to force through their strategic and operational decisions but also allow them to be the driving force in the negotiation with their supply chain partners (Benton and Maloni 2005; Cachon and Kök 2010; Touboulic et al. 2014). To manage supply chains strategically and operationally, it is essential for the management to understand the nature of power structure that exists in the supply chains (Cox 1999). Although power issues were widely explored in the operations and supply chain management literature (Cox 1999; 2001; 2004; Kadiyali et al. 2000; Benton and Maloni 2005; Raju and Zhang 2005; Pan et al. 2010; Kolay and Shaffer 2013; Touboulic et al. 2014), very few studies focus on the effect of power structure on operational decisions and supply chain performance in the context of service supply chains. Therefore, this book chapter aims to address this gap by exploring the following two research questions:

- How does the power dynamic in the service supply chain affect firms' operations decisions?
- What effect do different power structures have on the performance of individual firms and the service supply chain as a whole?

To answer these questions, we present two case studies. The first case focuses on a retail service supply chain, in which the mixture of offline and online channels adds a new dimension of competition. The second case focuses on a mobile phone service supply chain, which is a fast growing industry with increasing complexity in competition and dynamic power relationship. Both the retailing industry and the mobile telecommunication industry make an important contribution to the service sector in both developed and developing economies. With the rapid technological advance, there are great opportunities as well as challenges in both industries. There are also intense competitions which are often driven by some dominant players. The power structure has a significant impact on the management of these service supply chains. We employ the game theoretical approach to analytically examine the impact of different power structures on supply chain decisions and performances in both case studies. Our research findings lead to many interesting managerial insights, which will be beneficial for supply chain organizations to make important operational and strategical decisions and improve their competitiveness.

The rest of the chapter is organized as follow. Relevant literature regarding to the effect of power structure on supply chain management is reviewed in Section 7.2. Then the two case studies are presented in Section 7.3 and Section 7.4 respectively. Finally, concluding marks and possible future research directions are provided in Section 7.5.

7.2 LITERATURE REVIEW

The effects of power structure on supply chain decisions and performance have been recognized as an important topic in operations and supply chain management. To highlight our contribution, the literature that is representative and closely relevant to our study is reviewed in this section.

Studies on the effects of supply chain power relationship on various aspects of supply chain management have been widely reported in the literature. Among them, Choi (1991) is one of the earlier studies, which investigated the power relationship in a supply chain consisting of two manufacturers and a common retailer and its impacts on equilibrium prices and profits. Through studying the case where the supplier has dominant bargaining power and the case where the buyer is in a more dominant position, Ertek and Griffin (2002) analyzed the impact of power structure on price, profits and the market price sensitivity in a two-stage supply chain. Through their investigation on the impact of additional Internet channel on a company's stock market return, Geyskens et al. (2002) found that powerful companies with a few direct channels achieve more financial benefit than less powerful ones with a broader direct channel offering. Benton and Maloni (2005) empirically examined the influences of supply chain power on supplier satisfaction and their research demonstrated that it was essential to include power-satisfaction variable in any examination of supply chain partnerships. Geylani et al. (2007) proposed a theoretical model in a triadic supply chain setting involving two rival retailers with imbalanced market power in order to illustrate a manufacturer strategic response to the dominant retailer. Zhao et al. (2008) investigated the relationship between power, relationship commitment and the supply chain integration through an empirical survey of 617 manufacturing firms in China. Their analysis results indicated that different types of customer power affect manufacturer' relationship commitment in various ways. More recently, from a power perspective, Touboulic et al. (2014) demonstrated the power influences on how supply chain members manage their relationships and its effect of organizational response to the implementation of sustainability.

Among the existing studies on the effect of power structure on supply chain decisions and performance, the game theoretical approach seems to a popular method. For instance, Cai et al. (2009), from game theoretical perspectives, examined the effect of price discounts and pricing schemes on the dual-channel supply chain competition under different power structures. Cai (2010) investigated channel selection and channel coordination problem in a dual-channel supply chain from retailer Stackelberg, supplier Stackelberg, and Nash game theoretic perspectives. In addition, Zhang et al. (2012) also applied the game theoretical approach to analyze the impact of products' substitutability and channel position on pricing decisions in two dual-exclusive channels. Through a comparison between three different power structures, they found that a balanced power structure, in this case, the vertical Nash game delivered the best performance for the whole supply chain. Similarly, Shi et al. (2013) studied the impacts of different power structures including manufacturer and retailer Stackelberg games and Nash game on supply chains with a random and price-dependent demand. They found that while power structure has an influence on supply chain efficiency and such as influence depends on both expected demand and demand shock. Nevertheless, most of above studies focus on the vertical supply chain power relationship and very few studies consider the multiple dimensional power structures in their research. Wu et al. (2012) is one of very few studies that both the vertical and horizontal power structures are incorporated in the investigation on the pricing decisions in a supply chain that is consisted of one common supplier and two retailers.

Despite the growing interest among the supply chain scholars in studying the effect of power structure on supply chain decisions and performances, there is lack of research in the existing literature focusing on service supply chains. This research aims to address this gap by presenting the two case studies on the effect of power structure in the context of a retail service supply chain and a mobile phone service supply chain.

7.3 CASE STUDY 1: A RETAIL SERVICE SUPPLY CHAIN WITH A MIXED DUAL-CHANNEL

The retail sector is an important part for most developing and developed economies across the world. The rapid technological development, in particularly the extensive use of Internet and smart phones, has made a profound impact on how consumers buying goods and consuming retail services. We have witnessed a significant growth of online sales in the last two decades. Many high street retailers have invested heavily on online stores in order to share a slice of this significant sales growth. Nevertheless, adding an online sales channel to the existing offline channel imposes challenges for many retailers. For instance, how to set up a pricing policy in an online and offline mixed dual-channel to optimize retailer's profit, how do provide the retail service of each channel, and what is the impact of the power position that the retailer has in this supply chain on the pricing decisions. This case studies a two-echelon retail supply chain that is consisted of one retailer and one supplier. The retailer purchases goods from the supplier and then sells to its customers through both the online and offline channels. We investigate the effect of supply chain power structure on pricing decisions and supply chain performance.

The model development and analysis presented in the following sections are extended from the study of Chen et al. (2015). The model can be described as figure 7.1. The parameters and variables used in the models are defined as the notations shown in Table 7.1. Readers can refer to Chen et al. (2015) for the detailed mathematical proofs of propositions and lemmas. In addition, some key model assumptions are made as follows:

- To avoid trivialities, suppose $w > c$ and $p_i > w + c_i$ $(i = 1, 2)$. The first condition ensures profit for supplier, and the second condition ensures profit for both channels.

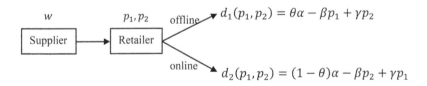

Figure 7.1 The Model Illustration.

- Without loss of generality, we assume that $c_1 > c_2$, which means that the offline unit sale cost is higher than that of online.
- We assume that the consumer demand via offline channel is $d_1(p_1,p_2) = \theta\alpha - \beta p_1 + \gamma p_2$ and the consumer demand via online channel is $d_2(p_1,p_2) = (1 - \theta)\alpha - \beta p_2 + \gamma p_1$, $\beta > \gamma$ (Mukhopadhyay et al. 2008; Chen et al. 2012; Wang and Li 2012). $\beta > \gamma$ means that the self-price sensitive is higher than the cross-price sensitive.
- It is assumed that both the retailer and the supplier are rational and self-interested. Each of them aims to maximize its own profit.

7.3.1 Retailer's pricing decision

The retailer has two options regarding its pricing decision: integrated pricing policy and decentralized pricing policy. Integrated pricing policy means that the retailer makes a centralized pricing decision for its online and offline channels. In contrast, for the decentralized pricing policy, the retailer makes the pricing decision for its online and offline channels separately.

7.3.1.1 Integrated pricing policy

For the integrated pricing policy, the objective of the retailer is to maximize the overall profit of offline and online channels. The retailer's overall profit from the two channels, denoted by $\pi_r(p_1,p_2)$, is

$$\pi_r(p_1,p_2) = (p_1 - w - c_1)d_1(p_1,p_2) + (p_2 - w - c_2)d_2(p_1,p_2) \tag{7.1}$$

The first term means the retailer's profit from the offline channel and the second term refers to the retailer's profit from the online channel. That is,

$$\pi_r(p_1,p_2) = (p_1 - w - c_1)(\theta\alpha - \beta p_1 + \gamma p_2) + (p_2 - w - c_2)[(1 - \theta)\alpha - \beta p_2 + \gamma p_1] \tag{7.2}$$

Table 7.1 Notations for Case Study 1.

Notation	Descriptions
c	Supplier's unit product cost
w	Supplier's unit wholesale price
p_1	Offline unit retail price
c_1	Offline unit sale cost
p_2	Online unit retail price
c_2	Online unit sale cost
α	The potential market scale
θ	The offline market share. $1 - \theta$ is the online market share
β	The self-price sensitive
γ	The cross-price sensitive
$d_1(p_1,p_2)$	Consumer demand via offline channel
$d_2(p_1,p_2)$	Consumer demand via online channel

As to the retailer's optimal price for offline channel (p_1^u) and online channel (p_2^u) in the integrated pricing policy, we obtain Lemma 7.1.

Lemma 7.1. $p_1^u = \frac{[\theta\beta+(1-\theta)\gamma]\alpha+(\beta^2-\gamma^2)(w+c_1)}{2(\beta^2-\gamma^2)}$ and $p_2^u = \frac{[(1-\theta)\beta+\theta\gamma]\alpha+(\beta^2-\gamma^2)(w+c_2)}{2(\beta^2-\gamma^2)}$.

7.3.1.2 Decentralized pricing policy

We defined the decentralized pricing policy as the scenario that the pricing decisions in the offline and online channel are made separately to maximize own channel's profits.

For the decentralized pricing policy, the offline channel's profit, demoted by $\pi_{r1}(p_1)$, is expressed as:

$$\pi_{r1}(p_1) = (p_1 - w - c_1)(\theta\alpha - \beta p_1 + \gamma p_2) \tag{7.3}$$

Similarly, the online channel's profit, denoted by $\pi_{r2}(p_2)$, is described as:

$$\pi_{r2}(p_2) = (p_2 - w - c_2)[(1-\theta)\alpha - \beta p_2 + \gamma p_1] \tag{7.4}$$

As to the retailer's optimal price for offline channel (p_1^s) and online channel (p_2^s) in the decentralized pricing policy, we obtain Lemma 7.2.

Lemma 7.2. $p_1^s = \frac{[2\theta\beta+(1-\theta)\gamma]\alpha+\beta\gamma(w+c_2)+2\beta^2(w+c_1)}{4\beta^2-\gamma^2}$ and $p_2^s = \frac{[\theta\gamma+2(1-\theta)\beta]\alpha+\beta\gamma(w+c_1)+2\beta^2(w+c_2)}{4\beta^2-\gamma^2}$.

7.3.1.3 The choice of pricing policy

In this section, we discuss the choice of retailer's pricing policy. Considering the maximum of $\pi_r(p_1, p_2)$, Lemma 7.1 and Lemma 7.2, the following proposition is obtained.

Proposition 7.1. $\pi_r(p_1^u, p_2^u) > \pi_r(p_1^s, p_2^s)$.

It means that the retailer's profit using the integrated pricing policy is higher than that using the decentralized pricing policy. Obviously, the retailer would prefer the integrated pricing policy. Therefore, it is assumed that the retailer adopts the integrated pricing policy when we model the pricing decisions under different power structures in the following sections.

7.3.2 Supplier Stackelberg model

In the Supplier Stackelberg power structure, the retailer and the supplier make their decision in sequence. The order of events is described as follows. Firstly, the wholesale price is announced by the supplier. Then, the retailer determines the offline and online retail prices according to the supplier's wholesale price. Third, the supplier chooses its optimal wholesale price using the response function of the retailer in order to maximize its profit. Finally, the supplier and the retailer gain their revenues when consumer demand is realized.

The supplier's profit in the Supplier Stackelberg power structure, denoted by $\pi_s(w)$, is described as:

$$\pi_s(w) = (w - c)d_1(p_1, p_2) + (w - c)d_2(p_1, p_2) \tag{7.5}$$

The first term represents the supplier's profit from retailer's offline channel sales and the second term is the supplier's profit from retailer's online channel sales. Equation (7.5) can be further expressed as:

$$\pi_s(w) = (w - c)[\alpha - (\beta - \gamma)(p_1 + p_2)] \tag{7.6}$$

As to the supplier's optimal price (w^s), retailer's optimal offline retail price (p_1^s) and optimal online retail price (p_2^s), we derive the following proposition.

Proposition 7.2. *In the Supplier Stackelberg Power Structure,* $w^s = \frac{\alpha}{4(\beta - \gamma)} + \frac{2c - c_1 - c_2}{4}$, $p_1^s = \frac{(1 + 4\theta)\beta + (5 - 4\theta)\gamma}{8(\beta^2 - \gamma^2)}\alpha + \frac{2c + 3c_1 - c_2}{8}$ *and* $p_2^s = \frac{(5 - 4\theta)\beta + (1 + 4\theta)\gamma}{8(\beta^2 - \gamma^2)}\alpha + \frac{2c - c_1 + 3c_2}{8}$.

It means that in the Supplier Stackelberg power structure, there exist unique supplier's optimal wholesale price and retailer's optimal offline and online retail prices.

From Equation (7.2) and Proposition 7.2, the retailer's maximum profit in the Supplier Stackelberg power structure is obtained as:

$$\pi_r(p_1^s, p_2^s) = \frac{16(\beta - \gamma)\theta^2 - 16(\beta - \gamma)\theta + 5\beta - 3\gamma}{32(\beta^2 - \gamma^2)}\alpha^2 - \frac{2c + c_1(8\theta - 3) + c_2(5 - 8\theta)}{16}\alpha$$
$$+ \frac{c(c + c_1 + c_2)(\beta - \gamma)}{8} + \frac{\beta[5c_1^2 - 6c_1 c_2 + 5c_2^2] + \gamma(3c_1^2 - 10c_1 c_2 + 3c_2^2)}{32} \tag{7.7}$$

From Equation (7.6) and Proposition 7.2, the supplier's maximum profit in the Supplier Stackelberg power structure is obtained as

$$\pi_s(w^s) = \frac{[\alpha - (\beta - \gamma)(2c + c_1 + c_2)]^2}{16(\beta - \gamma)} \tag{7.8}$$

From Equation (7.7) and Equation (7.8), the supply chain's profit in the Supplier Stackelberg power structure, denoted by $\pi(w^s, p_1^s, p_2^s)$, is obtained as

$$\pi(w^s, p_1^s, p_2^s) = \pi_r(p_1^s, p_2^s) + \pi_s(w^s) \tag{7.9}$$

7.3.3 Retailer Stackelberg model

In the Retailer Stackelberg power structure, the retailer and the supplier make their decisions in sequence. The order of events is described as follows. Firstly, the offline retail price and online retail price are announced by the retailer. Then, supplier decides its wholesale price according to the retailer's offline and online retail prices. Third, the retailer determines its optimal offline and online retail prices using the response function of the supplier in order to maximize its profit. Finally, the supplier and the retailer receive their revenues when the consumer demand is realized.

It is assumed that the marginal profit from the offline channel (denoted by m_1) is $m_1 = p_1 - w$, the marginal profit from the online channel (denoted by m_2) is $m_2 = p_2 - w$. Then, the supplier's profit, denoted by $\pi_s(w)$, is described as

$$\pi_s(w) = (w - c)d_1(p_1, p_2) + (w - c)d_2(p_1, p_2) \tag{7.10}$$

The first term represents the supplier's profit from retailer's offline channel sales and the second term is the supplier's profit from retailer's online channel sales. Equation (7.10) can be further expressed as:

$$\pi_s(w) = (w - c)[\alpha - (\beta - \gamma)(m_1 + m_2 + 2w)] \tag{7.11}$$

As to the supplier's optimal price (w^r), retailer's optimal offline retail price (p_1^r) and optimal online retail price (p_2^r), we obtain the following proposition.

Proposition 7.3. *In the Retailer Stackelberg power structure,* $w^r = \frac{\alpha}{8(\beta - \gamma)} + \frac{6c - c_1 - c_2}{8}$, $p_1^r = \frac{(1+4\theta)\beta + (5-4\theta)\gamma}{8(\beta^2 - \gamma^2)}\alpha + \frac{2c + 3c_1 - c_2}{8}$ *and* $p_2^r = \frac{(5-4\theta)\beta + (1+4\theta)\gamma}{8(\beta^2 - \gamma^2)}\alpha + \frac{2c - c_1 + 3c_2}{8}$.

This proposition means that in the Retailer Stackelberg power structure, there exist unique supplier's optimal wholesale price and retailer's optimal offline and online retail prices.

From Equation (7.2) and Proposition 7.3, the retailer's maximum profit in the Retailer Stackelberg power structure is derived as:

$$\pi_r(p_1^r, p_2^r) = \frac{8(\beta - \gamma)\theta^2 - 8(\beta - \gamma)\theta + 3\beta - \gamma}{16(\beta^2 - \gamma^2)}\alpha^2 - \frac{2c + c_1(4\theta - 1) + c_2(3 - 4\theta)}{8}\alpha$$
$$+ \frac{c(c + c_1 + c_2)(\beta - \gamma)}{4} + \frac{\beta[3c_1^2 - 2c_1c_2 + 3c_2^2] + \gamma(c_1^2 - 6c_1c_2 + c_2^2)}{16} \tag{7.12}$$

From Equation (7.6) and Proposition 7.3, the supplier's maximum profit Retailer Stackelberg power structure is obtained as

$$\pi_s(w^r) = \frac{[\alpha - (\beta - \gamma)(2c + c_1 + c_2)]^2}{32(\beta - \gamma)} \tag{7.13}$$

From Equation (7.12) and (7.13), the supply chain's profit in the Retailer Stackelberg power structure, denoted by $\pi(w^s, p_1^s, p_2^s)$, is calculated as

$$\pi(w^r, p_1^r, p_2^r) = \pi_r(p_1^r, p_2^r) + \pi_s(w^r) \tag{7.14}$$

7.3.4 Vertical Nash model

In a Vertical Nash power structure, the sequence of events is described as follows. The retailer decides its offline retail price and online retail price to maximize its profit given the supplier's wholesale price, and the supplier decides its wholesale price simultaneously to maximize its profit given the retailer's offline retail price and online retail price. Finally, the retailer and the supplier receive their revenue when the consumer demand is realized.

As to the supplier's optimal price (w^v), retailer's optimal offline retail price (p_1^v) and optimal online retail price (p_2^v), we obtain the following proposition.

Proposition 7.4. *In a Vertical Nash power structure,* $w^v = \frac{\alpha}{6(\beta-\gamma)} + \frac{4c-c_1-c_2}{6}$, $p_1^v = \frac{(1+6\theta)\beta+(7-6\theta)\gamma}{12(\beta^2-\gamma^2)}\alpha + \frac{4c+5c_1-c_2}{12}$ *and* $p_2^v = \frac{(7-6\theta)\beta+(1+6\theta)\gamma}{12(\beta^2-\gamma^2)}\alpha + \frac{4c-c_1+5c_2}{12}$.

It represents that in the Vertical Nash power structure, there exist unique supplier's optimal wholesale price and retailer's optimal offline and online retail prices.

From Equation (7.2) and Proposition 7.4, the retailer's maximum profit in the Vertical Nash power structure is calculated as:

$$\pi_r(p_1^v,p_2^v) = \frac{36(\beta-\gamma)\theta^2 - 36(\beta-\gamma)\theta + 13\beta - 5\gamma}{72(\beta^2-\gamma^2)}\alpha^2 - \frac{8c+c_1(18\theta-5)+c_2(13-18\theta)}{36}\alpha$$
$$+ \frac{2c(c+c_1+c_2)(\beta-\gamma)}{9} + \frac{\beta[13c_1^2 - 10c_1c_2 + 13c_2^2] + \gamma(5c_1^2 - 26c_1c_2 + 5c_2^2)}{72}$$

$$(7.15)$$

From Equation (7.8) Proposition 7.4, the supplier's maximum profit in the Vertical Nash power structure is calculated as:

$$\pi_s(w^v) = \frac{[\alpha - (\beta-\gamma)(2c + c_1 + c_2)]^2}{18(\beta-\gamma)} \tag{7.16}$$

From Equation (7.15) and Equation (7.16), the supply chain's overall profit in the Vertical Nash power structure, denoted by $\pi(w^v,p_1^v,p_2^v)$, is obtained as:

$$\pi(w^v,p_1^v,p_2^v) = \pi_r(p_1^v,p_2^v) + \pi_s(w^v) \tag{7.17}$$

7.3.5 Effect of power structure on retail supply chain management

In this section, the effects of different supply chain power structures on the retail supply chain's decisions and performances are discussed. We start with the examination of the impact of power structure on the retailer supply chain's optimal price. The following proposition is derived by comparing the retailer's optimal offline and online retail prices and the supplier' optimal wholesale price under three different power structures:

Proposition 7.5. $p_1^s = p_1^r > p_1^v$, $p_2^s = p_2^r > p_2^v$ *and* $w^s > w^v > w^r$.

It means that the retailer's optimal offline and online retail prices in the Supplier Stackelberg power structure are equal to those in the Retailer Stackelberg power structure. In other words, Supplier Stackelberg or Retailer Stackelberg power structure makes no difference on the retailer's pricing decisions. Moreover, the retailer's optimal offline and online retail prices in the Supplier Stackelberg and Retailer Stackelberg power structures are higher than those in the Vertical Nash power structure. This is due to the reason that there is a more balanced supply chain power relationship between the retailer and the supplier in the Vertical Nash power structure, and the competitive nature of the relationship that drives the retail prices down. As a result, consumers are generally better off in the Vertical Nash power structure. The proposition also indicates that the supplier's optimal wholesale price is more sensitive to the power structure as the supplier get three different optimal wholesale prices under three different power structures. More specifically, the supplier will set a low wholesale price when it has less market power and in contrast, a high wholesale price will be set when it is in a more powerful position.

Now, we look at the impact of power structure on the retail supply chain's maximum profit. The following proposition is obtained by comparing the retailer's maximum profits, the supplier' maximum profits and the supply chain's maximum overall profits under three different power structures.

Proposition 7.6. $\pi_r(p_1^r, p_2^r) > \pi_r(p_1^v, p_2^v) > \pi_r(p_1^s, p_2^s)$, $\pi_s(w^s) > \pi_s(w^v) > \pi_s(w^r)$ and $\pi(w^s, p_1^s, p_2^s) = \pi(w^r, p_1^r, p_2^r) < \pi(w^v, p_1^v, p_2^v)$.

From the above proposition, it is clear that the retailer achieves highest profit in the Retailer Stackelberg power structure and lowest profit in the Supplier Stackelberg power structure. The retailer's profit in the Vertical Nash Power Structure is between those achieved in other two power structures. In contrast, the order of the supplier's profit in the three different power structures is in an opposite direction. Not surprisingly, a powerful position over its supply chain counterparties will enable the company to gain more benefit. However, from the whole supply chain's perspective, a more balanced power relationship, the Vertical Nash power structure in this case, will deliver the best supply chain performance.

7.4   CASE STUDY 2: MOBILE PHONE SUPPLY CHAIN MANAGEMENT

The mobile phone sector has experienced continuous significant growth over the last two decades as mobile communications have become an irreplaceable part of people's lives in every continent of the world. Despite its importance, it is also one of the fastest growing sectors yet to receive much academic attention. At the same time, there is no let-up in technological advancements, product variety, shortened product life cycles, customer demand, and supply uncertainty, which have intensified competition in the industry (Cricelli et al. 2011). Interestingly, the mobile phone sector is one of the most competitive markets, in which we have witnessed some dramatic rises and falls of high profile companies. In this case study, a mobile phone supply chain is considered that consists of a handset manufacturer and a telecom service operator. This case study

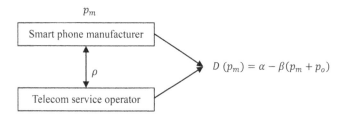

Figure 7.2 The Mobile Phone Supply Chain.

Table 7.2 Notations for Case Study 2.

Notation	Descriptions
p_m	Smart phone manufacturer's unit smart phone retail price
c_m	Smart phone manufacturer's unit handset manufacturing cost
p_o	Telecom service operator's unit service price
c_o	Telecom service operator's unit service cost
ρ	Smart phone manufacturer's unit subsidy from the telecom service operator
α	The primary demand of the mobile phone handsets
β	Customers' sensitivity to the mobile phone retail price

investigates the effect of different power structures on firms' decisions on pricing, subsidy and channel selection in the mobile phone supply chain.

The model development and analysis presented in the following sections are extended from the study of Chen and Wang (2015). The mobile phone supply chain models investigated in this research can be shown by figure 7.2. The parameters and variables used in the models are defined as the notations shown in Table 7.2. Readers can refer to Chen and Wang (2015) for the detailed mathematical proofs of propositions and lemmas. Moreover, some key model assumptions are made as follows:

- We consider the problem of channel selection in a two-echelon supply chain consisting of an upstream mobile phone manufacturer and a downstream telecom service operator in both a free channel and bundled channel. A free channel is defined as the setting in which consumers buy the handset from the manufacturer and the telecom service from a service operator separately. A bundled channel is defined as the setting, in which a mobile phone handset is designed to work exclusively with a single telecom service operator and consumers can only buy the bundled handset and the service package from either the manufacturer or the operator.
- In a free channel, the mobile phone manufacturer only has only one source of revenue, which is from the mobile phone handset sales. In a bundled channel, the manufacturer's revenue includes the handset sales and the subsidy ρ ($-p_m < \rho < p_o$) from the telecom service operator. Meanwhile, the telecom service operator's revenue includes the service revenue and the subsidy ρ ($-p_m < \rho < p_o$) from the mobile phone manufacturer. When the value of subsidy ρ is negative, it becomes the additional cost for the manufacturer or the service operator.

- Here, the customer demand is assumed to be linear and deterministic, which is defined as $D(p_m) = \alpha - \beta(p_m + p_o)$. This demand function has been widely employed in operations management and marketing research (Ingene et al. 1995; Padmanabhan et al. 1997; Chen et al. 2012).
- It is assumed that the telecom service operator's unit price (p_o) to be fixed at first. It is common in the mobile phone industry that the service price is relatively stable compared to the prices of mobile phone handsets. Nevertheless, we will analyze the effect of service unit price (p_m) on supply chain decisions in the later sections.
- We assume that both the mobile phone manufacturer and the telecom service operator are rational and self-interested. They both aim to maximize their own profits.

7.4.1 Free channel models

When a free channel is chosen for the mobile phone supply chain, the sequence of events is described as follows. First, the manufacturer determines the retail price for the mobile phone products according to the customer demand. When the customer demand is realized, the manufacturer receives sales revenue and the service operator receives service revenue.

The mobile phone manufacturer's profit in a free channel, denoted by $\pi_m^f(p_m)$, is obtained as

$$\pi_m^f(p_m) = (p_m - c_m)[\alpha - \beta(p_m + p_o)] \tag{7.18}$$

The telecom service operator's profit in a free channel, denoted by $\pi_o^f(p_m)$, is obtained as

$$\pi_o^f(p_m) = (p_o - c_o)[\alpha - \beta(p_m + p_o)] \tag{7.19}$$

As to the mobile phone manufacturer's optimal retail price in a free channel, denoted by p_m^f, the following proposition is derived:

Proposition 7.7. *In a free channel,* $p_m^f = \frac{\alpha}{2\beta} + \frac{c_m}{2} - \frac{p_o}{2}$.

This proposition means that there exists a unique optimal retail price for the mobile phone manufacturer in a free channel. From Proposition 7.7, we know that p_m^f is an increasing function of α and c_m, and is a decreasing function of β and p_o. It means that in a free channel, the manufacturer will set a high retail price if the primary demand of mobile phone handsets and the unit manufacturing cost are high. In contrast, the manufacturer will set a low retail price if the customers' sensitivity to the retail price of mobile phone products and the telecom operator's service price is high.

From Equation (7.18) and Proposition 7.7, the mobile phone manufacturer's maximum profit in a free channel is obtained as

$$\pi_m^f(p_m^f) = \beta\left(\frac{\alpha}{2\beta} - \frac{c_m}{2} - \frac{p_o}{2}\right)^2 \tag{7.20}$$

From Equation (7.19) and Proposition 7.7, the telecom service operator's maximum profit in a free channel is obtained as

$$\pi_o^f(p_m^f) = \beta(p_o - c_o)\left(\frac{\alpha}{2\beta} - \frac{c_m}{2} - \frac{p_o}{2}\right) \tag{7.21}$$

7.4.2   Bundled channel models

In a bundled channel, in addition to the products sales, the mobile phone manufacturer can gain additional revenue if they receive subsidy from the telecom service operator or generate cost if they give subsidy to the telecom service operator. Similarly, the service operator can add subsidy to service sales revenue if they receive subsidy from the manufacturer or generate additional cost if they give subsidy to the manufacturer. Therefore, the mobile phone manufacturer's profit in a bundled channel, denoted by $\pi_m^b(p_m)$, is expressed as

$$\pi_m^b(p_m) = (p_m + \rho - c_m)[\alpha - \beta(p_m + p_o)] \tag{7.22}$$

The telecom service operator's profit in a bundled channel, denoted by $\pi_o^b(\rho)$, is described as

$$\pi_o^b(\rho) = (p_o - \rho - c_o)[\alpha - \beta(p_m + p_o)] \tag{7.23}$$

Assume the mobile phone manufacturer's marginal profit in a bundled channel is m, then $m = p_m + \rho - c_m$, that is, $p_m = m + c_m - \rho$. By replacing p_m in (7.23), the telecom service operator's profit in a bundled channel ($\pi_o^b(\rho)$) can also be calculated as

$$\pi_o^b(\rho) = (p_o - \rho - c_o)[\alpha - \beta(m + c_m - \rho + p_o)] \tag{7.24}$$

7.4.2.1   Telecom service Operator Stackelberg model

In the Operator Stackelberg power structure, the mobile phone manufacturer and telecom service operator make their decisions in sequence. The order of events is described as follows. First, the manufacturer determines the retail price of mobile phone handset given the telecom service operator's subsidy. Then, the operator decides an optimal subsidy using the response function of the manufacturer to maximize its profit. Finally, the manufacturer and the operator receive their revenues when the customer demand is realized.

As to the mobile phone manufacturer's optimal retail price (p_m^t) and the telecom service operator's optimal subsidy (ρ^t) in a bundled channel with the Operator Stackelberg power structure, the following proposition is derived:

Proposition 7.8. *In a bundled channel with the Operator Stackelberg power structure, $p_m^t = \frac{3\alpha}{4\beta} + \frac{c_m}{4} + \frac{c_o}{4} - p_o$ and $\rho^t = p_o + \frac{c_m}{2} - \frac{c_o}{2} - \frac{\alpha}{2\beta}$.*

It means that in a bundled channel with the Operator Stackelberg power structure, there are unique optimal retail price for the handset and optimal subsidy between the manufacturer and the operator.

From Proposition 7.8, we can know that p_m^t is an increasing function of α, c_m and c_o, and is a decreasing function of β and p_o. ρ^t is an increasing function of p_o, c_m and β, and is a decreasing function of c_o and α. It means that in a bundled channel with the Operator Stackelberg power structure, the mobile phone manufacturer will set a high retail price if the primary demand of mobile phone handsets, the unit manufacturing cost, and the telecom service price are high. In contrast, the manufacturer will set a low retail price if customers' sensitivity to the handset retail price and the telecom service price are high. In addition, the operator would give more subsidies to the manufacturer if the telecom service price, unit manufacturing cost, and customers' sensitivity to the handset retail price are high. In contrast, the operator would give fewer subsidies to the manufacturer if the service cost and primary demand of handset are high.

From Equation (7.22) and Proposition 7.8, the mobile phone manufacturer's maximum profit in a bundled channel with the Operator Stackelberg power structure is obtained as

$$\pi_m^b(p_m^t) = \beta \left(\frac{\alpha}{4\beta} - \frac{c_m}{4} - \frac{c_o}{4} \right)^2 \tag{7.25}$$

From Equation (7.23) and Proposition 7.8, the telecom service operator's maximum profit in a bundled channel with the Operator Stackelberg power structure is obtained as

$$\pi_o^b(\rho^t) = 2\beta \left(\frac{\alpha}{4\beta} - \frac{c_m}{4} - \frac{c_o}{4} \right)^2 \tag{7.26}$$

Then, the maximum profit of the whole mobile phone supply chain in a bundled channel with the Operator Stackelberg power structure, denoted by $\pi^b(p_m^t, \rho^t)$, is obtained as

$$\pi^b(p_m^t, \rho^t) = 3\beta \left(\frac{\alpha}{4\beta} - \frac{c_m}{4} - \frac{c_o}{4} \right)^2 \tag{7.27}$$

7.4.2.2 *Vertical Nash model*

In the Vertical Nash power structure, the mobile phone manufacturer and telecom service operator make their decisions simultaneously. The order of events is described as follows. The manufacturer determines the retail price of mobile phone handset to maximize profit given the telecom service operator's subsidy, and the operator determines the subsidy to maximize profit given the retail price of handset. Finally, the manufacturer and the operator receive their revenues when the customer demand is realized.

As to the mobile phone manufacturer's optimal retail price (p_m^v) and the telecom service operator's optimal subsidy (ρ^v) in a bundled channel with the Vertical Nash power structure, we derive the following proposition:

Proposition 7.9. *In a bundled channel with the Vertical Nash power structure,* $p_m^v = \frac{2\alpha}{3\beta} + \frac{c_m}{3} + \frac{c_o}{3} - p_o$ *and* $\rho^v = p_o + \frac{c_m}{3} - \frac{2c_o}{3} - \frac{\alpha}{3\beta}$.

It means that in a bundled channel with the Vertical Nash power structure, there are unique optimal retail price for the handset and optimal subsidy between the manufacturer and the operator. From Proposition 7.9, we can know that p_m^v is an increasing function of α, c_m and c_o, and is a decreasing function of β and p_o. ρ^v is an increasing function of p_o, c_m and β, and is a decreasing function of c_o and α.

From Equation (7.22) and Proposition 7.9, the mobile phone manufacturer's maximum profit in a bundled channel with the Vertical Nash power structure is derived as

$$\pi_m^b(p_m^v) = \beta \left(\frac{\alpha}{3\beta} - \frac{c_m}{3} - \frac{c_o}{3} \right)^2 \tag{7.28}$$

From Equation (7.23) and Proposition 7.9, the telecom service operator's maximum profit in a bundled channel with the Vertical Nash power structure is derived as

$$\pi_o^b(\rho^v) = \beta \left(\frac{\alpha}{3\beta} - \frac{c_m}{3} - \frac{c_o}{3} \right)^2 \tag{7.29}$$

Then, the maximum profit of the whole mobile phone supply chain in a bundled channel with the Vertical Nash power structure, denoted by $\pi^b(p_m^v, \rho^v)$, is obtained as

$$\pi^b(p_m^v, \rho^v) = 2\beta \left(\frac{\alpha}{3\beta} - \frac{c_m}{3} - \frac{c_o}{3} \right)^2 \tag{7.30}$$

7.4.2.3 Manufacturer Stackelberg (MS) model

In the Manufacturer Stackelberg power structure, the telecom service operator and the mobile phone manufacturer make their decisions in sequence. The order of events is described as follows. First, the operator determines the subsidy given the manufacturer's retail price of handset. Then, the manufacturer decides an optimal retail price using the response function of the operator to maximize profit. Finally, the manufacturer and the operator receive their revenues when the customer demand is realized.

As to the mobile phone manufacturer's optimal retail price (p_m^m) and the telecom service operator's optimal subsidy (ρ^m) in a bundled channel with the Manufacturer Stackelberg power structure, we obtain the following proposition:

Proposition 7.10. *In a bundled channel with the Manufacturer Stackelberg power structure, $p_m^m = \frac{3\alpha}{4\beta} + \frac{c_m}{4} + \frac{c_o}{4} - p_o$ and $\rho^m = p_o + \frac{c_m}{4} - \frac{3c_o}{4} - \frac{\alpha}{4\beta}$.*

It means that in a bundled channel with the Manufacturer Stackelberg power structure, there are unique optimal retail price for the handset and optimal subsidy between the manufacturer and the operator. From Proposition 7.10, we can know that p_m^m is an increasing function of α, c_m and c_o, and is a decreasing function of β and p_o. ρ^m is an increasing function of p_o, c_m and β, and is a decreasing function of c_o and α.

From Equation (7.22) and Proposition 7.10, the mobile phone manufacturer's maximum profit in a bundled channel with the Manufacturer Stackelberg power structure is derived as

$$\pi_m^b(p_m^m) = 2\beta \left(\frac{\alpha}{4\beta} - \frac{c_m}{4} - \frac{c_o}{4} \right)^2 \tag{7.31}$$

From Equation (7.23) and Proposition 7.10, the telecom service operator's maximum profit in a bundled channel with the Manufacturer Stackelberg power structure is derived as

$$\pi_o^b(\rho^m) = \beta \left(\frac{\alpha}{4\beta} - \frac{c_m}{4} - \frac{c_o}{4} \right)^2 \tag{7.32}$$

Then, the maximum profit of the whole mobile phone supply chain's in a bundled channel with the Manufacturer Stackelberg power structure, denoted by $\pi^b(p_m^m, \rho^m)$, is obtained as

$$\pi^b(p_m^m, \rho^m) = 3\beta \left(\frac{\alpha}{4\beta} - \frac{c_m}{4} - \frac{c_o}{4} \right)^2 \tag{7.33}$$

7.4.3 The effect of power structure on mobile phone service supply chain management

In this section, we start with the analysis of the effect of power structure on the mobile phone supply chain's profit in a bundled channel. Through a comparison of optimal profits obtained in the previous sections, the following proposition is obtained:

Proposition 7.11. $\pi_m^b(p_m^m) > \pi_m^b(p_m^v) > \pi_m^b(p_m^t)$, $\quad \pi_o^b(\rho^t) > \pi_o^b(\rho^v) > \pi_o^b(\rho^m)$ $\quad$ and $\pi^b(p_m^v, \rho^v) > \pi^b(p_m^t, \rho^t) = \pi^b(p_m^m, \rho^m)$.

It means that the mobile phone supply chain member with more power will gain higher profit. In addition, the whole mobile phone supply chain achieves the highest maximum profit in a bundled channel with the Vertical Nash power structure. This finding is in line with many other studies that a balanced power structure is always the best for the whole supply chain (Zhang et al. 2012).

Furthermore, we discuss the channel selection policies for mobile phone supply chain. By comparing the optimal solutions and performances between the free channel and the bundled channel, we derive the following proposition:

Proposition 7.12. *(1) With the Operator Stackelberg power structure, if $\frac{\alpha}{2p_o + c_m - c_o} <$ $\beta < \frac{\alpha}{\frac{4}{3}p_o + c_m - \frac{1}{3}c_o}$, then the mobile phone supply chain will choose a bundled channel; if $0 < \beta \le \frac{\alpha}{2p_o + c_m - c_o}$, then the mobile phone supply chain will choose a free channel.*

(2) With the Vertical Nash power structure, the mobile phone supply chain will choose a free channel.

(3) With the Manufacturer Stackelberg power structure, if $\frac{\alpha}{(2+\sqrt{2})p_o + c_m + (1+\sqrt{2})c_o} <$ $\beta < \frac{\alpha}{2(2+\sqrt{2})p_o + c_m - (3+2\sqrt{2})c_o}$, then the mobile phone supply chain will choose a bundled

channel; if $0 < \beta \leq \frac{\alpha}{(2+\sqrt{2})p_o + c_m + (1+\sqrt{2})c_o}$, then the mobile phone supply chain will choose a free channel.

From this proposition, the optimal channel selection policies for a mobile phone supply chain can be obtained under different power structures. More specifically, a bundled channel will be chosen in both the Operator Stackelbeg and the Manufacturer Stackelberg power structures when customers' sensitivity to the retail price of mobile phone handset meets a certain condition. In contrast, a free channel will be selected in the Vertical Nash power structure.

7.5 CONCLUSIONS

This research investigates the impact of power structure on operational decisions and performances in the service supply chain. We present two case studies in the context of a retail service supply chain with the offline and online dual channels, and a mobile phone service supply chain with the free and bundled dual channels. The game theoretical approaches are applied to analytically examine the impact of different power structures on the operational decisions on pricing and channel selections and the performances of induvial firms and the whole supply chain in both case studies. Our analysis provides some interesting managerial insights that will help firms to make important strategic and operational decisions and improve their performances.

In the first case study, a retail service supply chain is considered, in which the retailer purchases products from the supplier and sells them to consumers through both offline and online retail channels. Through modelling the retail service supply chain with offline and online mixed dual channels, we derive many interesting findings. For example, our findings show that the retailer prefers the integrated pricing policy over the decentralized pricing policy. Furthermore, the supply chain power structure has a significant impact on the retailer and supplier pricing decisions and their performances individually and collectively. While individual supply chain members are better off when they have more supply chain power, but the whole supply chain will perform better if there is a balanced power distribution.

In the second case study, a mobile phone supply chain is considered that consists of a mobile phone manufacturer and a telecom service operator. We investigate the pricing, subsidy and channel selection decisions under different power structures. This research provides several interesting insights, which will enable the mobile phone manufacturer and the telecom service operator to develop appropriate pricing and subsidy policies to maximize their benefits. The research findings will also help them to formulate proper channel selection policies under different power structures to maximize their economic performance.

Similar to many other studies in the literature, there are some limitations of this research. Addressing these limitations will point to many interesting further research directions. For example, in both case studies, it is assumed that the retail service supply chain is composed by a suppler and a retailer, and the mobile phone service supply chain is consisted of one mobile phone manufacturer and one telecom service operator. One future research direction is to extend to multiple suppliers/manufacturers and retailers/service providers (Adida and Demiguel 2011; Pal et al. 2012). Furthermore, the demand functions adopted in both case studies are assumed to be linear deterministic demand. Although such demand functions are widely adopted in the supply chain literature, we have to acknowledge the uncertainty nature of demand, which may have an impact on the research outcome (He and Zhao 2012; Liu et al. 2013). One important extension of this work is to include stochastic demand in the modelling. Finally, service level should play an important role when we measure the performance of the service supply chain, which is one key difference compared with the conventional manufacturing supply chain (Schmitt 2011; Rezapour and Farahani 2014). Another possible research extension is to examine the impacts of service level on the service supply chain's decisions. It will be a more challenging research if these extensions are incorporated in the model.

REFERENCES

Adida, E., & DeMiguel, V. (2011). Supply chain competition with multiple manufacturers and retailers. *Operations Research*, 59(1), 156–172.

Benton, W.C., & Maloni, M., (2005). The influence of power driven buyer/seller relationships on supply chain satisfaction. *Journal of Operations Management*, 23(1): 1–22.

Cachon, G. P., & Kök, A. G. (2010). Competing manufacturers in a retail supply chain: On contractual form and coordination. *Management Science*, 56(3), 571–589.

Cai, G. (2010). Channel selection and coordination in dual-channel supply chains. *Journal of Retailing*, 86(1), 22–36.

Cai, G.G., Zhang, Z.G., & Zhang, M. (2009) Game theoretical perspectives on dual-channel supply chain competition with price discounts and pricing schemes. *International Journal of Production Economics*, 117(1): 80–96.

Chen, X., Ling, L., & Zhou, M. (2012). Manufacturer's pricing strategy for supply chain with warranty period-dependent demand. *Omega*, 40(12), 807–816.

Chen, X., Wang, X., & Jiang, X., (2015) The impact of power structure on retail service supply chain with an O2O mixed channel, *Journal of Operational Research Society*. DOI:10.1057/jors.2015.6

Chen, X. & Wang, X. (2015) Free or bundled: channel selection decisions under different power structures. *Omega*, 53, 11–20.

Choi, S. C. (1991). Price competition in a channel structure with a common retailer. *Marketing Science*, 10(4), 271–296.

Cox A. (1999) Power, value and supply chain management. *Supply Chain Management: An International Journal*, 4(4), 167–75.

Cox, A. (2001). Understanding buyer and supplier power: a framework for procurement and supply competence. *Journal of Supply Chain Management*, 37(1), 8–15.

Cox, A. (2004). The art of the possible: relationship management in power regimes and supply chains. *Supply Chain Management: An International Journal*, 9(5), 346–356.

Cricelli, L. Grimaldi, M., & Ghiron, N.L., (2011). The competition among mobile network operators in the telecommunication supply chain. *International Journal of Production Economics*, 131(1), 22–29.

Ertek G., & Griffin P.M. (2012) Supplier-and buyer-driven channels in a two-stage supply chain. *IIE Transactions*, 34(8), 691–700.

Geylani, T., Dukes, A. J., & Srinivasan, K. (2007). Strategic manufacturer response to a dominant retailer. *Marketing Science*, 26(2), 164–178.

Geyskens, I., Gielens, K., & Dekimpe, M. G. (2002). The market valuation of internet channel additions. *Journal of Marketing*, 66(2), 102–119.

He, Y., & Zhao, X. (2012). Coordination in multi-echelon supply chain under supply and demand uncertainty. *International Journal of Production Economics*, 139(1), 106–115.

Ingene, C.A. & Parry, M.E. (1995) Channel coordination when retailers compete. *Marketing Science*, 14(4), 360–77.

Kadiyali, V., Chintagunta, P., & Vilcassim, N. (2000). Manufacturer-retailer channel interactions and implications for channel power: An empirical investigation of pricing in a local market. *Marketing Science*, 19(2): 127–48.

Kolay, S. & Shaffer, G. (2013). Contract Design with a Dominant Retailer and a Competitive Fringe. *Management Science*, 59(9), 2111–2116.

Liu, W. H., Xu, X. C., & Kouhpaenejad, A. (2013). Deterministic approach to the fairest revenue-sharing coefficient in logistics service supply chain under the stochastic demand condition. *Computers & Industrial Engineering*, 66(1), 41–52.

Mukhopadhyay, S. K., Yao, D., & Yue, Q. (2008). Information sharing of value-adding retailer in a mixed channel hi-tech supply chain. *Journal of Business Research*, 61(9): 950–958.

Padmanabhan V. & Png, I.P.L. (1997) Manufacturers return policies and retail competition. *Marketing Science*, 16(1), 81–94.

Pal, B., Sana, S. S., & Chaudhuri, K. (2012). A three layer multi-item production–inventory model for multiple suppliers and retailers. *Economic Modelling*, 29(6), 2704–2710.

Pan, K., Lai, K. K., Leung, S. C., & Xiao, D. (2010). Revenue-sharing versus wholesale price mechanisms under different channel power structures. *European Journal of Operational Research*, 203(2): 532–538.

Raju, J., & Zhang, Z.J. (2005) Channel coordination in the presence of a dominant retailer. *Marketing Science*, 24(2): 254–62.

Rezapour, S., & Farahani, R. Z. (2014). Supply chain network design under oligopolistic price and service level competition with foresight. *Computers & Industrial Engineering*, 72, 129–142.

Schmitt, A. J. (2011). Strategies for customer service level protection under multi-echelon supply chain disruption risk. *Transportation Research Part B: Methodological*, 45(8), 1266–1283.

Shi, R., Zhang, J., & Ru, J. (2013). Impacts of Power Structure on Supply Chains with Uncertain Demand. *Production and Operations Management*, 22(5), 1232–1249.

Spring, M. & Araujo, L., (2009) Service, services and products: Rethinking operations strategy. *International Journal of Operations & Production Management*, 29(5), 444–467.

Stavrulaki, E. & Davis, M., (2014) A typology for service supply chains and its implications for strategic decisions. *Service Science*, 6(1), 34–46.

Touboulic, A., Chicksand, D., & Walker, H. (2014). Managing Imbalanced Supply Chain Relationships for Sustainability: A Power Perspective. *Decision Sciences*, 45(4), 577–619.

Wang, X., & Li, D. (2012). A dynamic product quality evaluation based pricing model for perishable food supply chains. *Omega*, 40(6), 906–917.

Wang, Y., Wallace, S. W., Shen, B., & Choi, T. M. (2015). Service supply chain management: A review of operational models. *European Journal of Operational Research*, 247(3), 685–698.

Wu, C. H., Chen, C. W., & Hsieh, C. C. (2012). Competitive pricing decisions in a two-echelon supply chain with horizontal and vertical competition. *International Journal of Production Economics*, 135(1), 265–274.

Zhang, R., Liu, B., & Wang, W. (2012). Pricing decisions in a dual channels system with different power structures. *Economic Modelling*, 29(2), 523–533.

Zhao, X., Huo, B., Flynn, B. B., & Yeung, J. H. Y. (2008). The impact of power and relationship commitment on the integration between manufacturers and customers in a supply chain. *Journal of Operations Management*, 26(3), 368–388.

Chapter 8

Resilience and complexity in a maritime service supply chain's everyday operation

Gesa Praetorius & Aditi Kataria

Maritime Risk and System Safety (MaRiSa) Research Group,
World Maritime University, Malmö, Sweden

SUMMARY

The maritime transport system is one of the major means for transporting goods as safely, efficiently and environmentally friendly as possible. In this system, ports represent hubs connecting maritime to other transport modes such as railway, road and aviation. Therefore ports and their services can be considered as critical bottlenecks where operations need to be sustained in a large variety of operating conditions. This chapter presents an analysis of the Vessel Traffic Service (VTS), a maritime shore-based information service which is part of a port's infrastructure. VTS is a key service in guaranteeing safe, fluent and efficient traffic flows in and out of a port, therefore making it a crucial part of a port's ability to provide a constant service performance despite the large uncertainties that are inherent to maritime operations. Functional Resonance Analysis Method (FRAM) is used to build a functional model of the VTS to analyse the system design and its impact on the service supply chain's ability to operate in a resilient manner, i.e. being able to sustain required operations prior, during and after disturbances or changes of operating conditions. While the chapter focuses on the maritime domain, it also provides an example on how a function-based approach can be used to understand and design service supply chains with a focus on how they achieve successful adaption to the large variety of operating conditions.

Keywords

Maritime transport system, functional resonance analysis method (FRAM), vessel traffic service, port operations

8.1 INTRODUCTION

In the 21st century global neo-liberal trade order, the world economies are inextricably interlinked in a complex transnational web of supply chains, irrevocably interconnecting the producing and consuming world regions (Bonacich & Wilson, 2008). The maritime transportation system is a vital component of worldwide multimodal transport by transporting nearly 90% of the world's trade (UNCTAD, 2013; George, 2013). It underpins the world economy by virtue of the sheer volume of cargo carried by sea and it is timely to consider the maritime transport from the perspective of Service

Supply Chain Management (SSCM); specifically ports, which have the potential to connect multiple transportation modes such as the railway, maritime and airborne transportation, and are an important component of the global value chain (Wang & Chen, 2010).

In recent years, the concept of resilience and resilience engineering has increasingly gained popularity in domains that provide important services to the general public, such as healthcare, nuclear power and aviation domains. Research within resilience engineering is concerned with how complex interdependent systems manage to sustain required operations under anticipated and unanticipated conditions (Hollnagel et al., 2011). Resilience engineering thus widens the perspective of the concept of supply chain resilience (Christopher, 2010; Falasca et al., 2008) from being the ability to recover from disturbances and returning to the originally desired state, to resilience being a characteristic of a system's performance that is designed to sustain functioning in a large variety of conditions, further acknowledging the need for performance to be variable to provide the capability to the system to absorb disturbance, to coordinate and relocate processes and resources in a flexible manner to create a buffering capacity which can come into play in situations where the system operates close to the edges of its performance envelope.

This chapter will present an analysis of the Vessel Traffic Service (VTS), a maritime shore-based information service which is part of a port's infrastructure. VTS is a key service in guaranteeing safe, fluent and efficient traffic flows in and out of port, therefore making it a crucial part of a port's ability to provide a constant service performance despite the large uncertainties that are inherent to maritime operations. Functional Resonance Analysis Method (FRAM) is used to build a functional model of the VTS to analyse the system design and its impact on the service supply chain's ability to operate in a resilient manner. While the chapter focuses on the maritime domain, it also provides an example on how a function-based approach can be used to understand and design service supply chains with a focus on how they achieve successful adaption to the large variety of operating conditions encountered in real-life settings.

8.2 SERVICE SUPPLY CHAIN MANAGEMENT AND THE MARITIME TRANSPORTATION DOMAIN

Supply Chain Management (SCM) is defined as the *"management of upstream and downstream relationships with suppliers and customers to deliver a superior customer value"* (Christopher, 2005, p. 5). Maritime transport can be identified as a supply chain as goods are shipped across the globe with service considerations that concern the safe loading, carriage and unloading of cargo (Stopford, 2009; HMSO, 1992). The overall goal of the transportation system is to ship the goods as safely and efficiently as possible, with ports providing the main connection between maritime transport and other transportation modes. Thus, the port and its services to the merchant fleet can be considered as a Service Supply Chain System (SSCS), a system which connects customers (merchant vessels, cargo owners) with suppliers (port services such as VTS, pilotage, tug service, berthing service etc) within a provided infrastructure (i.e. berths, port approaches) and where the product, by its nature is the safe, fluent and efficient transport and handling of cargo. Wang et al. (2015) distinguish two specific types of

SSCS; Service Only Supply Chains (SOCS) and Product Service Supply Chains (PSSC). SOCSs are systems where the product is defined as pure service, while PSSCs are systems that produce both a physical product and provide services. Examples for SOCSs are telecommunication and financial services, whereas examples for PSSCs are healthcare and customer services. As a port provides a physical infrastructure as well as services, it can be considered as a PSSC. Mapping the concept of a SSCS onto a port can help to appreciate its pertinence to the domain in terms of the identification of the diverse suppliers, service providers, connected units and consumers that together form a network (see Baltacioglu et al., 2007). Maritime transport is unlike a traditional retail space. The latter allows for a straightforward identification of the three parties namely, infrastructure service provider, retail service provider and customer, that compose an SSCS as identified by Demirkan and Cheng (2008). The analogy of a service supply chain system is extendable to the maritime transport system, however it needs to be noted that several SSCS can exist in the overall maritime SSC from the very first supplier in the chain to the final customer (Ellram et al., 2004).

8.2.1 Vessel Traffic Service – A maritime information service system

Vessel Traffic Service (VTS) is a shore-sided service within a country's territorial waters with the aim to support the fluent, safe and efficient flow of traffic within a determined area (VTS area). VTS Operators (VTSOs) monitor the traffic, assist in navigational matters, and provide information to all vessels in the service centre's area of responsibility; normally this corresponds to port areas or areas that pose navigational difficulties for the ship's bridge team (IALA, 2012).

VTS can be delivered on three different service levels: Information Service (INS), Traffic Organisation Service (TOS), and Navigational Assistance Service (NAS). INS constitutes broadcasting information to all merchant vessels within the VTS area on a specific VTS radio channel. It may contain information relevant for the safe passage of the area, and can consist of reports on position, identity and intentions of other traffic, or information concerning the meteorological and geographical state of the area. TOS is the operational management of traffic movements conducted through Very High Frequency (VHF) broadcasts. It aims to prevent the development of dangerous situations as well as to avoid congestion. NAS is an intervention in the decision making on board with the aim to assist the traffic in a safe and expedient passage by providing information on the VHF. Ships are offered instructions only when safety is at risk or upon their request as the decision making power within maritime operations remains on the bridge of a vessel. As the international framework identifies VTS as an assistance service, VTSOs do not in general have the legal mandate to actively manage maritime traffic through tasks such as voyage optimisation, route planning, or the planning of traffic density in the area.

Hitherto, most research about the VTS has been conducted within the area of equipment development with a strong emphasis on mathematical models and information fusion (i.e. Chang, 2004; Kharchenko & Vasylyev, 2004; Vespe et al., 2008). Furthermore, several studies have stressed the importance to understand the interaction between operator and technology (i.e. Brodje et al., 2013; Nuutinen et al., 2006), and the need to understand VTS as a domain for traffic management (van Westrenen,

2014, van Westrenen & Praetorius, 2014). However, VTS has not yet been addressed as part of the maritime SSCS. Regardless the scarcity of research so far, VTS should be of interest to SSCM as the port is a key transitional hub in intermodal/multimodal transport, where the modality changes from seaborne to land-based rail and road transport to penetrate and reach consumers in the hinterland.

8.3 VTS AS A SERVICE ONLY SUPPLY CHAIN

The VTS essentially offers services and thus can be classified as a Service Only Supply Chain (SOSC) where the product is itself a service that's provided, as defined by Wang et al. (2015). The authors (ibid, p. 686) utilise three core concepts of service supply chains, *service supply management, service demand management and service supply chain coordination*. Through these three concepts applied to the VTS, it becomes salient that SOCS and service supply chain management can offer a valuable perspective to analyse the VTS.

- *Service supply management* – as with other SOSCs, the vertical supply chain in a VTS is short with the service directly provided by the VTS to the vessels in the delineated VTS area. However, similar to other SOSCs, the horizontal service supply chain for the VTS is long. The complexity of the horizontal service supply chain is exemplified by the diverse stakeholders involved in VTS such as the shipping companies, their local agents, port authorities, pilots, tugs, docks etc. The key issue facing the VTS in terms of SSCM is disruption risk. Disruption risks to the VTS can be exemplified by maritime accidents that can throw marine traffic out of gear, disrupt service provision and have large attendant economic costs. Disruption risks may also be posed by marine traffic that arrives late and misses the tidal window to enter the port. Service competition is largely felt by the port authority which can lose revenue to another port due to its lower port dues and permission to sail with pilot exemption certificates instead of compulsory pilotage.
- *Service demand management* – the demand for the VTS services largely comes from the vessels in the VTS area. The Marine traffic calling at the port may require pilotage services, lighterage, information pertaining to the transit and information on the organisation and ordering of traffic, if any, till the vessel is safely alongside. Forecasting this demand; planning for it while taking into account the uncertainty is crucial to the demand management of the VTS service. The key issues with respect to service demand management in the VTS are service capacity and supply management with disruption risk and service. Depending upon the additional functions performed by the VTS, it might need to arrange and/or provide information with respect to the availability of pilots, berths etc. at this juncture as the VTS is crucial as the link between the port and the ship.
- *Service supply chain coordination* – the length of the horizontal service supply chain of the VTS involves, but is not limited to several participants such as the port authority, pilots, tugs, docks, shipping companies, local agents and shipboard seafarers, among others. Coordination and collaboration between these members of the VTS service supply chain is imperative to its performance which is primarily

to ensure the safe and efficient movement of traffic in the waters under its purview. The key issues facing the VTS service supply chain can be identified as the provision of customer service, management of service supply with disruption and risk and service.

8.3.1 Resilience engineering and supply chain management

In the domain of supply chain management, resilience is associated with the ability to absorb disturbances and regain the originally desired state (Christopher, 2010). It has been emphasized that this ability is crucial for supply chains to be able to quickly respond to disruptions and mitigate their consequences. One of the key features in supply chain management resilience is the sharing of information as today's SCMs are identified as complex networks with multiple upstream and downstream dependencies among the various nodes within the chain.

Resilience Engineering (RE) is a relatively young body of research that emerged at the beginning of the 2000s. Resilience, which has its origin as a concept within ecology in the early 1970s, defines an ecological system's ability to arrive at an equilibrium, or stable state, over time in a dynamic and changing environment (Holling, 1973). In the context of large-scale systems, such as the maritime transportation domain, resilience is the ability to sustain required functioning and achieve system goals in the light of anticipated and unanticipated conditions. Resilience engineering thus becomes the theoretical framework concerned with the design of complex systems that succeed to operate in a large variety of operational conditions and match challenges, including disturbances that arise through the dynamic character of the environment (Praetorius et al., 2015).

In RE, systems are analysed with the aid of four cornerstones – monitoring, response, anticipation, and learning, which characterise the features a system should have to be able to maintain its functioning before, during and after anticipated and unanticipated events have occurred. The focus is on a system's successful adaption of performance to the demands within the environment (Hollnagel et al., 2006). When adaption is successful, safety and efficiency emerge as a property, as the system balances goals and demands in the current context (Woods, 2006), e.g. safe and efficient traffic movement within a port approach. The four cornerstones can be used to understand a system's performance and provide insights in how resilience manifests itself in the everyday production along the service supply chain. Further, by analysing everyday operations with the aid of the abilities, one is able to identify ways in which the system's capacity for knowing what to do (respond), what to look for (monitor), what to expect (anticipate) and what has occurred (learn) can be strengthened (Hollnagel, 2011a). This can in turn inform design activities and help to identify ways to make a system, such as a SOCS, more bumpable (Woods & Hollnagel, 2006) in the sense that it will be able to operate under a variety of conditions without major performance drops.

8.4 FUNCTIONAL RESONANCE ANALYSIS METHOD

The FRAM is a method to analyse and model complex sociotechnical systems, in which functions are distributed over human operators, organisations and technology.

The method has its origins within accident and event analysis and has been developed in the early 2000s as a novel method to complement commonly used human reliability assessment methods (Hollnagel, 2004). The aim of the method is to model, analyse and understand system performance with a focus on the concept of performance variability and ways in which systems manage and monitor potential and actual variability. FRAM is based on four basic principles; the principle of equivalence of successes and failures, principle of approximate adjustments, principle of emergence and the principle of functional resonance (e.g. Herrera & Woltjer, 2010; Hollnagel, 2014). The *principle of equivalence of successes and failures* expresses that the only difference in between these two is the judgement of the outcome. While an action is deemed a success if it has the desired outcome, the same action can be identified as a failure when negative and unforeseen consequences occur. How these consequences arise is accounted for by the *principle of approximate adjustments*. Sociotechnical systems are complex systems acting in an uncertain and dynamic environment. Functions are distributed over people, technology and organisation that adjust their performance to be able to meet the demands the system is facing in the current situation. As this adjustment is based on the availability of resources (e.g. time, manpower), it will always be approximate. Consequently, everyday performance is, and needs to be variable to help the system to successfully adapt its functioning to the current operational conditions. While variability within one function possibly can be managed or monitored, the *principle of emergence* emphasizes that variability in multiple functions may combine in unanticipated ways and cause disproportional and non-linear effects. Although performance variability can lead to negative outcomes, it is first and foremost necessary for a system's resilience, for the ability to function under beneficial and harmful conditions alike. The last principle, *the principle of functional resonance*, highlights the potential of the variability in multiple functions to resonate, and therefore reinforce and even amplify itself, so that the outcome of a function might carry an unusually high amount of variability, which the system is not able to manage given the current condition. As a result, accidents might occur.

FRAM consists of four steps which are used to model the system based on functions and to identify sources of performance variability as well as measures to manage, dampen or monitor it. In *Step 1* all necessary system functions are defined. The aim is to afford a consistent description as a basis of the analysis. All functions are described in form of their six aspects (Input, Output, Time, Control, Precondition, Resources/Executing conditions, Table 8.1). These aspects describe the basic characteristic of an activity and help to understand relations among functional units within a system.

The functions that are the focus of the analysis are called foreground functions. Functions that are required by the foreground functions, but which do not themselves contribute to the variability being investigated, are called background functions (Hollnagel, 2012). Background functions represent the context and while they do not vary during the time frame specific for the analysis, they shape the performance and affect how events progress (Hollnagel et al., 2014).

Step 2 helps to identify the variability of the functions in the FRAM model. The functions' performance can vary in various ways. While functions involving humans tend to vary a lot, technical functions usually show a stable performance over time. Organisational functions do not show the same extent of variability as human

Table 8.1 Aspects of a function (adapted from Hollnagel, 2012).

Aspect	Description
Input (I)	Conventional input and/or a signal that activates the function, is used or transformed by the function (requires change of state for the function to start)
Output (O)	Result of what the function does, represents a change of the system's state or output parameters
Precondition (P)	Conditions that need to be fulfilled before the function can be carried out
Resource (R)/executing condition	Material or matter that are consumed, or executive conditions, that need to be present, while the function is active
Control (C)	Supervises or regulates the function so that it derives the desired output
Time (T)	Aspects of time that affect the way the function is carried out

functions, but show a delayed effect on these. There are three types of variability that can be characterized in a function: endogenous, exogenous, and upstream-downstream coupling variability. Endogenous variability arises due to the nature of the function and is therefore internal, while exogenous variability is due to the variability of external factors, such as the work environment. Most interesting for an event analysis is the upstream-downstream coupling variability as it can become the basis for functional resonance. Upstream functions are carried out before downstream functions in the instantiation of the model, which means that variability in the earlier will impact on the performance of the latter (Hollnagel, 2012).

In *Step 3* of the analysis, an instantiation is created to see how performance variability can propagate through the system. It can help to understand how performance variability within some functions can amplify or dampen the variability of other functions, as the instantiation provides a way of simulating the functions performance within a specific operational condition to identify vulnerabilities and strengths of the system at work. The final step, *Step 4*, is used to suggest ways in which performance variability can be monitored, managed or eliminated. Based on the results from step 2 and 3 it is analysed how to best monitor and manage the variability where it is necessary. However, the analyst should keep in mind that varying performance also is an indicator for the flexibility to adapt performance to specific conditions, which means that eliminating variability can make the system rather inflexible and brittle (Woods, 2006).

8.5 UNDERSTANDING EVERYDAY OPERATIONS AND ADAPTION IN THE VTS

To explore everyday operation in the VTS, a workplace study was conducted in the VTS of an international South Asian port. Data generation research techniques for the study involved interviews (n = 10), observations and audio recording of the communication between port, VTS, pilot service and merchant vessels on the VHF radio. Interview data was annotated and approximately 30 1/2 hours of audio of the VHF interaction, were transcribed verbatim and thematically coded to facilitate the analysis. For the purpose of this analysis the trajectories of inbound and outbound vessels from a port are differentiated to explore the coordination required to perform the requisite functions and provide the necessary services to the vessels.

8.5.1   Everyday operations as linear processes

The two main processes under analysis in this chapter are the provision of service to inbound vessels and outbound vessels. This section shows the two as linear processes and exemplifies how the vessels' way in and out of port can be understood as a trajectory. One way of depicting these trajectories, emphasizing the contribution of the VTS as SOCS, is utilising flow diagrams. The flow diagrams (Fig. 8.1, Fig. 8.2) have been derived through the identification of patterns revealed with the analysis of transcripts that included conversation analysis and further supplemented by observations and interviews. Figures 8.1 and 8.2 below depict the trajectories of inbound (Fig. 8.1) and outbound (Fig. 8.2) vessels from the port. The vessel trajectories are largely derived from the real time interaction between the VTS and the ships on the VHF radio augmented by interview findings and ethnographic observations.

An inbound vessel contacts the port VTS before entering the port limits and several times during the course of its journey through the channel. An analysis of the data reveals that inbound ships bound for the port first call the VTS from outside the port limits. Depending upon the day's docking programme/schedule, the VTSOs call the ships in ample time, giving the vessel adequate notice to prepare the vessel and set sail. Once under way the vessel could call the VTS upon entering port limits. Thereafter, the vessel calls again when nearing the pilot station. At this juncture the vessel also communicates with the pilot regarding the instructions for pilot boarding and subsequent to pilot boarding; the pilot takes on the VHF communication with the VTS until he disembarks after berthing or anchoring the vessel as the case may be. This completes the trajectory of a typical inbound vessel. The Only difference between a vessel that arrives without a pre-booked berth and one that has a prior booking is that the former is instructed to drop anchor at the designated anchorage until plans are put in place to call the vessel inside.

Figure 8.2 highlights the procedure and the process of managing an outbound vessel by the VTS. The process of initiating the procedure for processing an outbound vessel commences with the vessel contacting the VTS and giving its readiness to sail. Upon confirmation of the port clearance certificate on-board, the VTS passes on the vessel's particulars and request for pilotage to the office which allocates the pilot. Upon receipt of the confirmation from the office in the form of an updated traffic plan, the VTS confirms the pilotage to the vessel and accordingly it departs under pilotage at the allocated time. Taken together both figures 8.1 and 8.2 graphically represent the trajectories of these vessels and are linear in their outlook and only represent the interaction of individual vessels in their dealings with the VTS, so showing the contribution of the VTS to the supply chain as a whole. The VTS provides the service at every stage of the journey and is needed to accomplish the overall goal to facilitate safe, efficient and fluent traffic movements.

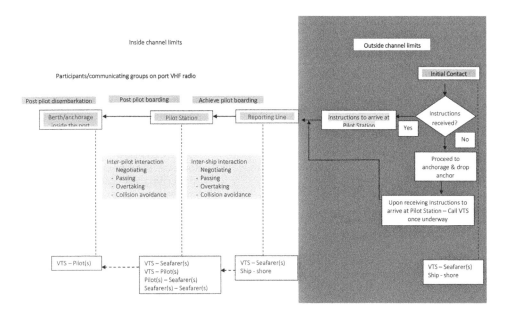

Figure 8.1 Trajectory of an inbound vessel.

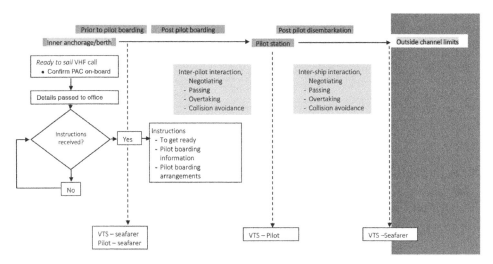

Figure 8.2 Trajectory of an outbound vessel.

8.5.2   Everyday operations through the lens of FRAM

While the above has shown the trajectory of inbound and outbound vessels, this section will show the overall complexity of everyday operations. FRAM has been applied to the same data set to analyse the functional design of the service supply chain system and identify how its design promotes resilience in the service provision.

Table 8.2 Identified functions and potential (output) variability.

Identified functions	Potential (output) variability
(to) establish contact with VTS	Too early, too late or omission
(to) report to VTS	Too early, too late or omission
(to) obtain vessel information	Imprecise, acceptable
(to) request pilotage information	Too early, too late or omission
(to) pass vessel information to office	Imprecise, acceptable
(to) give anchoring information	Too early, too late or omission
(to) anchor outside channel limits	Too early, too late or omission
(to) pass reporting point	Too fast, too slow
(to) enter channel	Too early, too late or omission
(to) update running logs	Imprecise, acceptable
(to) prepare traffic plan	Imprecise, acceptable
(to) give traffic plan	Too early, too late or omission
(to) receive traffic plan	Too early, too late or omission
(to) give updates to traffic plan	Imprecise, acceptable
(to) receive updated traffic plan	Too early, too late or omission
(to) call the ship for pilotage	Too early, too late or omission
(to) provide Information Service	Imprecise, acceptable, Too early, too late or omission
(to) monitor traffic	Imprecise, acceptable,
(to) provide Navigational Assistance Service	Imprecise, acceptable
(to) initiate port departure	Too early, too late or omission
(to) give readiness to sail	Too early, too late or omission
(to) confirm port clearance	Imprecise, acceptable
(to) inform ship about departure pilotage	Too early, too late or omission
(to) inform about traffic of neighbouring VTS	Imprecise, acceptable, too early, too late or omission
(to) receive traffic information from neighbouring VTS	Imprecise, acceptable, too early, too late or omission
(to) establish SOPs	Imprecise, acceptable

Step 1: Identification of the SSCS's functions

26 functions resembling everyday operation have been identified (table 8.2). Of these 26 functions, 3 are background functions ((to) establish contact with VTS, (to) establish SOPs, (to) inform about traffic of neighbouring VTS) and 23 foreground functions.

Step 2: Identification of variability

Table 8.2 depicts the 26 functions and their potential variability. Each of the function has the potential to introduce variability into the system's performance, where variability of several functions may give rise to functional resonance.

Within this model external variability is introduced to the functions through the large extent of uncertainty which arises through external – for example, limited time horizon with regards to how many vessels the VTS will have to handle, availability of pilotage and berths – and environmental, i.e. hydro meteorological conditions such as water levels, wind, current, factors beyond the control of the service itself.

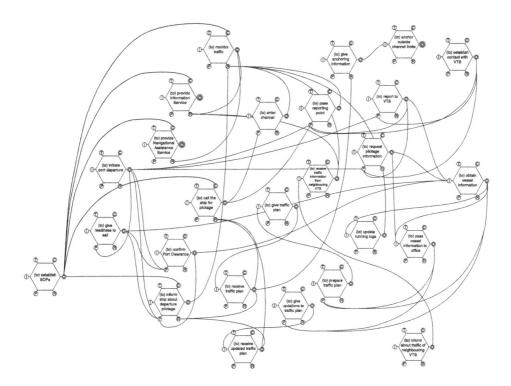

Figure 8.3 FRAM model of VTS functions.

Step 3: The functional model of the SOCS

Figure 8.3 depicts the functional model of the SOCS. It includes both the outbound and inbound trajectory as functions are largely the same for both processes. In the model the inherent complexity of the service provision becomes visible. While functions may be distributed over various organisations (VTS, port, ship, neighbouring VTS, they still are needed to accomplish the overall product of the VTS SOCS. The model in fig. 8.3 depicts all potential couplings between the functional units. Depending on a specific scenario, e.g. a vessel outbound from the port, a certain set of functions will be activated. Not all couplings need to be active at all times, however, it needs to be noted that each of the couplings between the functions is a potential path through which performance variability can spread and possibly accumulated to functional resonance. It is hence necessary to visualise all couplings to be able to identify which of the functions are the most critical and should be sustained in any operating condition.

Step 4: Identification of mitigation strategies

Within this study, step 3 has been concerned with the identification of all potential couplings among the functions to provide a visualisation of the complexity of the everyday operation of the VTS SOCS and its contribution to realisation of fluent, safe and efficient traffic movements. No specific instantiation dealing with a single event has been developed and hence the possibility to suggest and reason about measures to eliminate, manage, monitor, or mitigate the consequences of functional resonance are limited. However, the model clearly depicts the upstream and downstream relationships between the functions and can thus be used to identify which functions are critical for the SOCS to sustain required operations.

The four critical functions required to sustain VTS operations are – *(to) monitor traffic; (to) call the ship for pilotage; (to) inform ship about departure pilotage and (to) obtain vessel information*. These functions trigger relevant couplings which resonate through and have the potential for breakdown should resonance go unchecked. *(to) monitor traffic* is one of the most crucial functions and utilises the output of five other functions as resources (*(to) pass reporting point; (to) enter channel; (to) update running logs; (to) receive traffic information from neighbouring VTS*). Similarly, pilotage of both incoming and outgoing vessels is of crucial importance to keep traffic flowing smoothly and efficiently. The pilotage functions are underpinned by the functions related to the preparation, updating and sharing of the traffic plan which is prepared by the port office. The traffic plan in turn is underpinned by the information which is obtained from the vessels and passed along to the office in real time for the processing of vessels to commence. The variability in the functions has been identified in table 8.2. The overall resonance of variability in the 5 crucial functions identified above can cause potential incidents and have a detrimental effect in the sustenance of overall operation. The system absorbs and dampens the variability in 4 of the crucial functions (*(to) monitor traffic; (to) call the ship for pilotage; (to) inform ship about departure pilotage; (to) obtain vessel information*) by calling upon additional manpower as required. There are four VTS operators on duty in a 24 hour shift and additional VTS operator is drafted in times of increasing workload with respect to the increasing demand.

8.6   DISCUSSION

The result section of this chapter presents two different types of analysis, an ethnographic study drawing upon conversation analysis and a FRAM, of everyday VTS operations at a South Asian port. The flow diagrams (fig. 8.1, fig. 8.2) can accommodate one trajectory at a time, either inbound or outbound, thus identifying all necessary steps in VTS operations. While this can help to understand the way in which different organisations are coupled to each other to accomplish the trajectory along the SSCS, it does not help to understand the complex net of upstream and downstream relationships within this SOCS.

In contrast to the linear process model in the flow diagrams, the FRAM model (fig. 8.3) accommodates both trajectories in one model, along with the additional functions of receiving information from the neighbouring VTS pertaining to their traffic as well as the background functions related to the preparation, updating and receiving of the traffic plan from the port office that is used as a key resource for the work of the VTS. The FRAM analysis of the VTS system under study thus reveals the complex interactions that constitute everyday operations. The couplings within the model depict all upstream and downstream relationships among the functions distributed over supplier of the infrastructure (port including provided services) and the customer (merchant vessel). VTS is only one part of the whole chain and already shows a high potential for functional resonance as the functions constituting everyday work are highly interdependent, which can help variability in one or more functions to spread easily through the system and in worst case scenario, reinforce itself causing functional resonance, and maybe a breakdown which the system needs to recover from.

Christopher (2010) argues that successful risk management in supply chain systems amongst others depend on understanding and improving the supply chain system, identifying and managing the critical paths and improving the network visibility. The results above show that FRAM has the potential to be a tool that can help with these activities. Through the functional model (fig. 8.3), the understanding of the overall system interactions (couplings) as well as its network visibility increase as critical functions are identified and their potential variability is analysed. At one glance it is possible to see that the functions are highly interconnected and show a high degree of dependency on each other within the system to be able to successfully provide fluent and safe traffic movements within the VTS area. The FRAM model depicts everyday work and is thus a snapshot of everyday complexity, which can be used to identify critical paths and bottlenecks for the system's ability to attain to its system goal. For the purpose of risk assessment and management, such a model can be used to develop specific scenarios, such as used by Herrera et al. (2010), which in turn can be used to identify potential hazards and their consequences, potential mitigation strategies, as well as performance indicators for safe performance. This can particularly be important for the process of risk assessment and management, where decisions on how to avoid and reduce the potential for system breakdowns as well as on how to mitigate potential consequences of identified risks are made. FRAM offers a way of visualising, modelling and analysing the complexities inherent in everyday work. Within the modelling process, the couplings, potential exogenous and endogenous as well as coupling variability are identified. By following the paths of the functional model during specific scenarios, input on how variability spreads through the functions and how it may accumulate to functional resonance can be obtained. Specifically critical are the upstream-downstream relationships among functions and the way in which they are affected by variability in one or more functions within the model. While risk assessment and analysis normally aims at eliminating and controlling risks as far as reasonably possible (Hollnagel, 2014) based on generic operational models, the FRAM offers a chance to look into everyday work and the way it unfolds based on the trade-offs, such as safety and efficiency, that the system faces and how it accomplishes goals under a large variety of operating conditions. Scenario-based approaches with FRAM, such as presented in Herrera et al. (2010) and in Macchi, Hollnagel and Leonhardt (2009), show a promising novel approach to risk assessment and management as they go beyond traditional approaches and reason on how systems in high-risk environments manage limited resources and deal with trade-offs generated by conflicting system goals. In addition to analysing everyday work, FRAM also explicitly focuses on how to manage variability. As failure and success stem from the same source (Hollnagel, 2014), the difference to risk management is that the system's functioning is seen in context and that the basis to successful system performance and system failure are essentially the same. It is about how and where performance variability arises and how its spreads, meaning that rather than eliminating the sources of variability, which would be the case if a traditional assessment methodology, such as a Failure Mode and Effect Analysis (Mikulak et al., 2008), FRAM aims to provide a means on how to manage this variability. As variability is the core to why systems are able to adapt and cope with everyday complexity, the elimination of such variability would not make the system necessarily safer, but rather more brittle, which in turn will affect its ability to operate in a resilient manner and bounce back.

Furthermore, three core concepts are crucial to address in the analysis of SSCS (Wang et al., 2015); supply chain management, service demand management, and service supply chain coordination. While the trajectory-based analysis depicted through the flow diagrams can help to enrich the understanding for the overall SSCS and the overall way in which the system is organised, the FRAM can signify a contribution to deepen the understand of how the system manages shifts in demand and coordinates its functional setup to be able to respond to and deal with changes in the system's environment. SOCS are highly interdependent and due to the pressure for efficient operations, are also highly coupled. A functional model can help to understand these couplings as functions span across organisation (ship, VTS, pilot, port) to accomplish the overall system goal (fluent, safe and efficient traffic movements). This type of modelling everyday operations can help to understand how functions affect each other and where a closer cooperation and coordination might be needed to increase the overall system performance.

The FRAM model presented here is only the basic model depicting all interdependencies, but it has not yet been instantiated to represent the interactions of the function in a specific case. As FRAM can be used for predictive and retrospective analysis (Hollnagel, 2012), the model above (fig 8.3) has now the potential to be applied to a certain set of cases to understand in detail how performance variability will be used by the system to deal with a shift in demands and situations, create buffers when needed and respond to critical situations. While the application here has addressed the maritime domain, the authors believe that FRAM can lend itself to analyse a wide variety of SSCS to address the system's potential for resilience. Hitherto resilience has often been associated with the reduction of slack and the optimisation of processes using methods such as the Six Sigma (Christopher, 2010), a functional model can show that systems actively create slack to be prepared for the unanticipated and to be able to relocate resources in the face of system disturbances. While variability and complexity based on interdependences and upstream-downstream couplings can put a SSCS at risk, the variability will in the end also be what makes a system resilient, adaptable and bumpable so that it can sustain required operation despite large amounts of uncertainty.

8.7 CONCLUDING REMARKS

This book chapter presented the VTS as a SOCS with the help of two analytical tools to visualise everyday operation and gain insights in how the system coordinates across the SSC and adapts to current operating conditions. FRAM has been introduced as a promising alternative to current approaches as it can help the analyst to understand the everyday complexity within the system under study and show how the system adapts to and copes with this complexity. Identified system functions, their upstream-downstream couplings, and potential variability in the functions' performance provide a new angle on everyday operation and on how a system adapts. This can be crucial for the development of risk management strategies and can help to identify critical paths and mitigation strategies which are focused on how to support adaption and coping, rather than eliminating hypothetical hazards to increase a system's safety.

The notion of performance variability and the use of FRAM for the analysis and design of complex socio-technical system is still in its beginnings. During the last decade FRAM has mostly been applied as a tool for retrospective event analysis focusing on how and why functional resonance has occurred in a specific situation. However, it shows a high potential for risk assessment as systems today are becoming more inherently complex with a large degree of interdependencies, which gives rise to the need for novel methodological approaches that can show how systems cope with and adapt to the variety of operating conditions they meet. Thus there is the need for further research into how to best integrate functional modelling into risk assessment to identify successful management strategies. As a next step within this research, it is suggested to use the functional models developed within a risk assessment process, such as a Formal Safety Assessment (Psaraftis, 2012) to test their potential to complement current safety assessment methodologies and to develop performance indicators for resilient performance, i.e. the system's potential to absorb disturbances and maintain required operations prior to, during and after an event, in VTS systems.

As discussed in this chapter, functional models can serve as a source of information about critical paths designed into a SSCS as well as to increase the overall visibility of the complex interdependencies designed into the supply chain system. While there is a limited body of operational research addressing functional models, it is believed that these can provide important input on how a SSCS works in situ, what critical trade-offs the system faces, and how it copes with those. As mentioned above, there is the potential to develop performance indicators for the system's overall performance and analysing SSCSs with the help of FRAM can be a first step in this development. However, more research within SSCS and FRAM is needed to develop this initial work presented here.

Table of notations

FRAM	Functional Resonance Analysis Method
IMO	International Maritime Organization
INS, TOS, NAS	Information Service, Traffic Organization Service, Navigational Advice and Assistance Service; three service levels that can be provided by a VTS center
PSSC	Product Service Supply Chain
RE	Resilience Engineering
SCM	Supply Chain Management
SSCM	Service Supply Chain Management
SSCS	Service Supply Chain System
SOP	Standard Operating Procedure
SOSC	Service Only Supply Chain
VHF	Very High Frequency radio; main means of communication between maritime stakeholders involved in port operations such as merchant vessels, VTS, pilotage and tug services
VTS	Vessel Traffic Service; a maritime information service

REFERENCES

Baltacioglu, T., Ada, E., Kaplan, M. D., Yurt, O., & Kaplan, Y. C. (2007). A new framework for service supply chains. *The Service Industries Journal, 27*(2), 105–124.

Bonacich, E., & Wilson, J. B. (2008). *Getting the goods: ports, labor and the logistics revolution.* Ithaca: Cornell University Press.

Brodje, A., Lundh, M., Jenvald, J., & Dahlman, J. (2013). Exploring non-technical miscommunication in vessel traffic service operation. *Cognition, Technology & Work, 15*(3), 347–357.

Chang, S. J. (2004). *Development and Analysis of AIS Applications as an Efficient Tool for Vessel Traffic Service.* Paper presented at the OCEANS '04.

Christopher, M. (2010). *Logistics & Supply Chain Management*: Gardners Books.

Demirkan, H., & Cheng, H. (2008). The risk and information sharing of application services supply chain. *European Journal of Operational Research, 187*(3), 765–784.

Ellram, L. M., Tate, W. L., & Billington, C. (2004). Understanding Service Supply Chain Management. *The Journal of Supply Chain Management, 40*(4), 17–32.

Falasca, M., Zobel, C. W., & Cook, D. (2008). *A decision support framework to assess supply chain resilience.* Paper presented at the 5th International ISCRAM Conference, Washington, DC, USA.

George, R. (2013). *Ninety percent of everything: inside shipping, the invisible industry that puts clothes on your back, gas in your car, food on your plate* New York: Metropolitan Books.

Herrera, I. A., & Woltjer, R. (2010). Comparing a multi-linear (STEP) and systemic (FRAM) method for accident investigation. *Reliability Engineering and System Safety (RESS), 95*(12), 1269–1275.

HMSO. (1992). *Carriage of Goods by Sea Act 1992.* London: HMSO.

Holling, C. S. (1973). Resilience and Stability of Ecological Systems. *Annual Review of Ecology and Systematics 1973, 4*, 1–23.

Hollnagel, E. (2004). Barriers and Accident Prevention. Aldershot, UK: Ashgate.

Hollnagel, E. (2011). Prologue: The Scope of Resilience Engineering. In E. Hollnagel, J. Pariès, D. Woods, & J. Wreathall (Eds.), *Resilience Engineering in Practice. A Guidebook.* Farnham, Surrey, UK.: Ashgate Publishing.

Hollnagel, E. (2012). *FRAM: The Functional Resonance Analysis Method – Modelling Complex Socio-Technical Systems*: Ashgate Publishing Company.

Hollnagel, E. (2014). The Four Basic Principles of the FRAM. Retrieved from http://functionalresonance.com/basic-principles.html (20150930)

Hollnagel, E., Hounsgaard, J., & Colligan, L. (2014). *FRAM – the Functional Resonance Analysis Method – a handbook for the practical use of the method.* Middelfart: Centre for Quality.

Hollnagel, E., Pariès, J., Woods, D., & Wreathall, J. (Eds.). (2011). *Resilience Engineering in Practice. A Guidebook.* Farnham, Surrey, UK.: Ashgate Publishing.

Hollnagel, E., Woods, D. D., & Leveson, N. (2006). *Resilience Engineering: Concepts And Precepts*: Ashgate.

IALA. (2012). *IALA Vessel Traffic Manual 2012* (5 ed.): International Association of Marine Aids to Navigation and Lighthouse Authorities.

John, A., Riahi, D., Paraskevadakis, A., Bury, A., Yang, Z., & Wang, J. (2015). A new approach for evaluating the disruption risks of a seaport system. In T. Nowakowski, M. Mlynczak, A. Jodejko-Pietruczuk, & S. Werbinska-Wojciechowska (Eds.), *Safety and reliability: Methodology and applications* (pp. 591–598). London: Taylor & Francis.

Kharchenko, V., & Vasylyev, V. (2004). *Decision-Making System for Vessel Traffic Planning and Radar Control.* Paper presented at the European Radar Conference Amsterdam.

Macchi, L., Hollnagel, E., & Leonhardt, J. (2009). *Resilience Engineering approach to safety assessment: an application of FRAM for the MSAW system*. Paper presented at the EUROCONTROL Safety R&D Seminar, Germany.

Mikulak, R. J., McDermott, R., & Beauregard, M. (2008). *The Basics of FMEA, 2nd Edition*: CRC Press.

Nuutinen, M., Savioja, P., & Sonninen, S. (2006). Challenges of developing the complex socio-technical system: Realising the present, acknowledging the past, and envisaging the future of vessel traffic services. *Applied Ergonomics, 28*, 513–524.

Praetorius, G., Hollnagel, E., & Dahlman, J. (2015). Modelling Vessel Traffic Service to understand resilience in everyday operations. *Reliability Engineering & System Safety*, 141, 10–21.

Psaraftis, H. N. (2012). Formal Safety Assessment: an updated review. *Journal of Marine Science and Technology*, 17(3), 390–402.

Sacks, H. (1989). *Lectures on Conversation 1964–1965 (edited by G. Jefferson with introduction by E. A. Schegloff)*. London: Kluver Academic Publishers.

Sacks, H. (1992). *Lectures on conversation (2 Vol.ed)*. Edited by G. Jefferson. Introduction by E Schegloff. Oxford: Blackwell.

Stopford, M. (2009). *Maritime Economics*. London: Routledge.

UNCTAD. (2013). *Review of Maritime Transport*. Retrieved from Geneva:

van Westrenen, F. (2014). Modelling arrival control in a vessel traffic management system. *Cognition, Technology & Work*, 1–8.

van Westrenen, F., & Praetorius, G. (2014). Maritime traffic management: a need for central coordination? *Cognition, Technology & Work*, 16(1), 59–70.

Vespe, M., Sciotti, M., Burro, F., Battistello, G., & Sorge, S. (2008). *Decision Support Platforms for Satellite-Extended Vessel Traffic Services*. Paper presented at the RADAR.

Wang, J. J., & Cheng, M. C. (2010). From a hub port city to a global supply chain management center: a case study of Hong Kong. *Journal of Transport Geography*, 18(1), 104–115.

Wang, Y., Wallace, S. W., Shen, B., & Choi, T.-M. (2015). Service supply chain management: A review of operational models. *European Journal of Operational Research*, 247(3), 685–698.

Woods, D. (2006). Essential Characteristics of Resilience. In E. Hollnagel, D. Woods, & N. Leveson (Eds.), *Resilience Engineering: Precepts And Concepts* (pp. 21–34). Abingdon: Ashgate Publishing Group.

Woods, D. D., & Hollnagel, E. (2006). *Joint Cognitive Systems. Patterns in Cognitive Systems Engineering*. Boca Raton, FL: CRC Press.

Chapter 9

Fast fashion retail operations services: An empirical study from consumer perspectives[1]

Wing-Yan Li[a], *Tsan Ming Choi*[a] *& Pui-Sze Chow*[b]

[a]*Business Division, Institute of Textiles and Clothing, The Hong Kong Polytechnic University, Hung Hom, Kowloon, Hong Kong*
[b]*School of Management, Centennial College, Pokfulam, Hong Kong*

SUMMARY

We examine in this empirical study how fast fashion service influences consumer shopping behaviors. Following the literature, we first develop hypotheses and conduct a consumer survey. With the collected data, we perform statistical analysis and reveal how the quick response service, pricing and design strategies of fast fashion retailers affect the involvement, as well as other purchasing attitudes and behaviors of the consumers. Our analysis reveals that: A higher consumer satisfaction level on the fast fashion strategies leads to a higher degree of customer involvement, which also has a positive impact on the customer attitudes towards the fast fashion retailers. Practical implications from this study are discussed.

Keywords

fast fashion, service marketing, fashion retailing, fashion retail operations, service management

9.1   INTRODUCTION

"Fast fashion" is a business model in which the offered fashion collections are based on the most recent fashion trends, driven by the market need (Choi, 2013). Under this business model, all products are designed and manufactured promptly. The product selling price is usually affordable by the mass market. Generally speaking, a fast fashion system combines the following two components (Swinney and Cachon, 2011): (i) Short production and distribution lead times which would enable a close matching of supply and demand (i.e., achieving a quick response system), and (ii) highly fashionable product designs (i.e., enhancing product design).

The fast fashion business model operates by understanding the target market's requirements, gathering the latest fashion trend information then quickly producing the merchandise in a cost-effective manner. For distribution, the up-to-date fashion

---

[1]This paper is part of the graduation FYP thesis of the first author.

products are shipped to the retail stores quickly. This helps to attract consumers to stop by more frequently and enjoy the retail services. The above fast fashion operations strategy has been widely adopted by many well-established large-scale international fashion retailers such as H&M and Zara.

This research aims at (i) understanding how fast fashion companies attract consumers in terms of their pricing, design, and quick response services, (ii) investigating how these strategies influence the degree of consumer involvement, and (iii) figuring out how consumer involvement would further influence consumers' shopping, purchase and after-purchase attitudes and behaviors.

This research makes an important contribution: By knowing how consumer behavior is influenced by fast fashion services, a fashion retailer, which is also a fast fashion service provider, can learn more about how to target and satisfy its customers, and also improve its production and marketing strategies.

The rest of this paper is organized as follows. Section 9.2 reviews the related literature. Section 9.3 presents the research methodology. Section 9.4 describes the data analysis and presents the findings. Section 9.5 concludes this paper with discussions on the implications of research findings and future research.

9.2 LITERATURE REVIEW

9.2.1 Related literature

In recent years, many fashion retailers have the tendency to focus on improving the "speed to market" with their maximum ability to produce new fashion items rapidly according to the latest fashion shows so as to increase their competitive advantages among competitors. According to Barnes and Lea-Greenwood (2006), fashion retailers such as Zara, and H&M adopt such a fast fashion system to produce and deliver fashion apparel rapidly and introduce the latest runway designs to the stores within a few weeks so as to attract consumers. Fast fashion retailers emphasize on rapidly changing product lines, high fashion content, reasonable price and reasonable quality. They gain an advantage by reflecting the newness of its offerings and create a sense of scarcity, which can result in a positive word of mouth. Those fast fashion retailers create a sense of "buy now because you won't see this item later", reinforcing a climate of scarcity and opportunity by small shipments, sparsely stocked display, time limit of individual items could be sold in the stores and a degree of intentional undersupply. The concept of quick response is to draw consumers back to the store for frequent visits by creating very new and up-to-date fashion merchandise within a short period of time. Quick response also encourages the application of new and innovative technologies into production with a goal to increase efficiency (Bruce and Daly, 2006) By applying the system of quick response, Jackson (2001) argues that fashion retailers shift from trends-forecasting (a push process) to using real-time market data to comprehend the needs and desires of the consumers (a pull process). Fast fashion can achieve the consumer-driven approach. The process of forecasting and product planning mostly depends on the varying market demand. It is argued that fast fashion retailers make reference to the famous designs and styles from top fashion brands in fashion shows and fashion magazines (Christopher et al., 2004). As a remark, most of the fast fashion retailers (like Zara) have their own creative teams which consist of designers, sourcing specialists, and product development experts. These teams all work together very closely. To establish an initial collection, they create a number of variations based on the innovative designs from the observed fashion trends, and then expand upon successful product items to continue the in-season development. At the same time, the optimal selection of fabrics and offering of product mix are decided.

Pricing is critical in retailing as well as any other business service operations. Jobber (2007) states that many consumers use pricing as an indicator of quality and they assume that highly priced products would "naturally" have a higher quality level. Attwood (2007) claims that although customers like lower prices, retailers cannot just depend on offering the lowest prices to achieve a competitive edge. In fast fashion, retailers still use low price as one of their strategies. Undoubtedly, there is a huge market segment consisting of fashion-conscious customers who prefer fashionable clothes with low price and a reasonable quality level. Many of these customers also adopt the deposable fashion concept in which they will throw away the old merchandise and buy new ones quickly. The success of a fast fashion retailer is realized by a faster turnover of inventory. In addition, statistical analysis reveals that a faster stock turnover has a positive effect to business performance.

Nowadays, fashion consumers are much more demanding and fashion conscious than before. As information and trends moving around the world at a very high speed, consumers are exposed to exclusive designs and styles from fashion runways more quickly. As a consequence, they have more options and thus can shop more often (Hoffman, 2007). "Shopping behaviors" explain how and where a consumer shops (McKinney et al., 2004). The comfort level of purchasing, frequency and amount of time per shopping and the consumer spending for an outfit can be useful data to the fashion retailers. Consumers shop in various kinds of stores with differing frequencies and spend inconsistent sum of money for a wide variety of merchandise. In the literature, O'Cass (2000) proposes a model to measure customer involvement (CI), which consists of four critical elements, namely: products, purchase decision, advertising, and consumption. As a matter of fact, consumers who find values from a brand and treat it as an important thing to themselves will be more likely to develop good attitudes towards the brand.

9.2.2 The empirical model and hypotheses

According to Muncy and Hunt (1984), consumers have significant connection between the value system and CI. Consumers are more likely to take part once they recognize that certain objects or stimulus have the potential to fulfill their needs and desires (Zaichkowsky, 1985). Fast fashion concept is popular in cities like Hong Kong; thus its trend may interact with consumer's value system and influence the level of CI. Based on the reviewed literature, fast fashion retailers have three main strategies, which are pricing, designs of fashion products, and quick response supply chain operations model. Thus, Hypotheses 1a, 1b and 1c are proposed to help uncover the relationship between those characteristics of fast fashion and CI.

Hypothesis 1a. Pricing strategy has a significant effect on customer involvement.

Hypothesis 1b. Design strategy has a significant effect on customer involvement.

Hypothesis 1c. Quick response strategy has a significant effect on customer involvement.

Hypothesis 2 is set to further investigate the relationship between customer involvement and consumers' attitudes towards fast fashion. It is known that the level of customer involvement in fashion products or brands may influence the shopping, purchase and after-purchase attitudes and behaviors of the consumers. As explained by Slama and Tashchian (1985), the action of purchasing a particular product would affect a consumer's attitudes and behaviors associated with that particular product. In turn, consumers with a higher level of CI with an object (i.e. product or brand) would be expected to possess a more positive attitude towards the same object. Thus, the following hypothesis is set:

Hypothesis 2. Customer involvement has a significant effect on consumer attitudes towards fast fashion.

9.3 METHODOLOGY

This study adopts the approach of an empirical consumer survey and the respective statistical analysis. The main objective of this paper is to investigate the relationship between fast fashion strategies and customer involvement, and the subsequent impact of customer involvement to customers' attitudes towards fast fashion. All questions are set based on the findings from literature review. The questionnaire survey involved non-probability convenience sampling. The questionnaire started with the filtering statement to ensure that all respondents had prior experience in purchasing from a fast fashion brand, either Zara or H&M. The set of questionnaire was then further divided into 4 sections, each of which consists of statements related to a particular variable under investigation. Respondents were asked to rate their extent of agreement to individual statements under a 6-point Likert scale, from "1 – Strongly disagree" to "6 – Strongly agree". Section 9.1 included 12 statements that solicit consumer's perception of fast fashion brands in terms of pricing, product design, and quick response. Section 9.2 consisted of 9 statements which were related to customer involvement with and purchasing decision towards fast fashion brands. Section 9.3 consisted of 7 statements which helped figure out customers' attitudes and shopping behaviors towards fast fashion brands. Finally Section 9.4 included three questions aiming at collecting respondents' profiles. Overall, 200 valid questionnaires were collected.

9.4 EMPIRICAL DATA ANALYSIS AND FINDINGS

The profiles of the respondents are summarized in Figures 9.1(a)–9.1(c).

After collecting the data, we conduct reliability test to assess the internal consistency of the measurement scales. The value of the Cronbach's alpha of all five categories

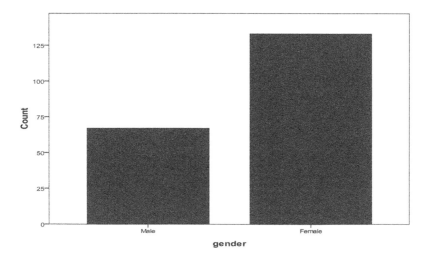

Figure 9.1(a) Gender of Respondents.

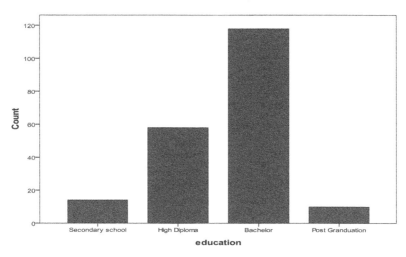

Figure 9.1(b) The Education Background of Respondents.

(pricing, design, quick response, customer involvement, and customers' attitude) are all over 0.7 (see Table 9.1). The reliability of the measurement items is validated.

9.4.1 Customer perception towards fast fashion and customer involvement

Correlation analysis is first applied to examine the relationship between fast fashion strategies and customer involvement (Table 9.2). The results of the correlation analysis show that the three fast fashion strategies, namely: "pricing", "design" and "quick response", are all positively correlated to customer involvement.

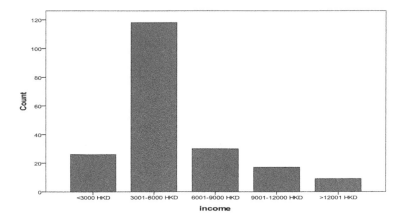

Figure 9.1(c) The Income Level of Respondents.

Table 9.3 presents the means score of customer perception towards each strategy. We discuss each one of them in the following.

(i) Pricing strategy

The statistical results show that respondents are quite satisfied with the pricing of fast fashion brands (mean score = 4.35) and the pricing strategy of fast fashion has a significant effect on customer involvement as the Pearson correlation coefficient is 0.913 (p-value = 0.000 < 0.01). To most of the customers (especially those who are price sensitive), price is a critically important factor in affecting their purchasing decision. In the literature, Josiam et al. (2005) argue that the more involved a consumer is, the more likely s/he is to shop for a longer period of time. If consumers perceive the price of a fashion product as reasonable and attractive, they will go shopping in those fashion brands more frequently and would be more actively involved in shopping there. Also, once a brand is perceived as affordable, customers would be more willing to spend and purchase more from it.

(ii) Design strategy

The result from Table 9.3 shows that respondents are satisfied with the design of fast fashion products (mean score = 4.46) and the design strategy of fast fashion brands has a significant effect on customer involvement as the Pearson correlation coefficient is 0.872 (p-value = 0.000 < 0.01). Most of the fast fashion retailers have their own design teams to work on product design and development. The teams would usually make reference to other famous designs and styles from fashion shows and prestigious fashion magazines, when creating their own designs. This strategy can help those fast fashion brands offer very timely and trendy products to attract those fashion-conscious customers. Observe that the fashion conscious customers want to find fashion products that suit their personality and images. They normally believe that the clothes that they wear will express their personal styles. Thus, the "enhanced design" strategy of fast

Table 9.1 Results of Reliability Tests of Various Variables.

Variable	Cronbach's alpha	Item	
Pricing	.922	1	The pricing of the merchandises sold in these retailers is reasonable.
		2	The pricing of their fast fashion products is within my spending budget.
		3	The pricing of these fast fashion products is more attractive with comparing with other brands.
		4	The pricing of these fast fashion products motivates my purchase.
Design	.906	1	I like the designs of the fashion products sold by these retailers as they are suitable to me.
		2	I like the designs of the fashion products sold by these retailers as they are very fashionable.
		3	I like the designs of the fashion products sold by these retailers as they are similar to those in high fashion brands.
		4	I like the various styles of fashion products in these stores.
Quick Response	.872	1	I like there is always new arrival of products in these stores, no matter the style is suitable to me or not.
		2	I like there is always new arrival of products in these stores, no matter how small is the new collection.
		3	The frequent new arrival of fast fashion products is significant to me.
		4	The frequent new arrival of products let me learn the latest trend.
Customer Involvement	.907	1	I am interested in fast fashion clothing.
		2	I pay a lot of attention to fast fashion clothing I spend lots of time finding fashion product I like.
		3	I am very excited to find out the fast fashion products in these stores.
		4	I visit these stores in every shopping trip.
		5	I visit these stores once I have spare time.
		6	I spend a relatively long time shopping these stores when comparing with other brands.
		7	The purchase decisions I make for fast fashion clothing are important to me.
		8	I think a lot about my purchase decisions when it comes to fast fashion.
		9	I like spending time on mix and match the fashion items I bought in these stores.
Customers' Attitudes	.860	1	I like shopping in these stores.
		2	They are my favorite stores.
		3	These stores are significant to me.
		4	My intension in purchasing in these stores is high.
		5	These stores provide products from different categories for my purchase.
		6	I purchase in these stores repeatedly.
		7	I would recommend products from these stores to my friends.

Table 9.2 The Summary of Correlation Analysis – Hypothesis 1.

Hypotheses	Pearson Correlation	Sig	Hypothesis supported/rejected
1a	0.913	0.00	supported
1b	0.872	0.00	supported
1c	0.920	0.00	supported

Table 9.3 The Mean Score of Fast Fashion Strategies.

Fast Fashion strategies	Mean score
Pricing	4.35
Design	4.46
Quick Response	4.48

fashion retailers can successfully increase customer involvement in terms of shopping time.

(iii)   Quick response

The result shows that respondents are quite satisfied with the quick response strategy of fast fashion brands (mean score $= 4.48$) and the quick response strategy of fast fashion has a significant effect on customer involvement as the Pearson correlation coefficient is 0.920 (p-value $= 0.000 < 0.01$). This finding is consistent with the literature. For instance, Cotte and Wood (2004) comment that customers with a higher degree of innovativeness would have a higher tendency to try new products/services willingly. They would buy new products more often (Roehrich, 2004). Such consumer behavior will be reinforced by the quick responsive strategy of fast fashion brands. Fast fashion retailers, like Zara and H&M, launch their new collections continuously within a very short period of time in a small quantity. This would better satisfy customer demand. In addition, customers are constantly changing their lifestyles due to socio-cultural factors and there is an increasing demand of self-uniqueness (Sproles and Burns, 1994). Quick response services stimulate customers to be more proactive in searching for fashion information. Frequent arrivals of new fashion products would further stimulate and excite these customers. Thus, they would visit those fast fashion brands more frequently to see whether there is any new product suitable for them, increasing customer involvement in terms of shopping frequency.

9.4.2   Leading strategies to customer involvement

To verify and compare the respective impact of the three fast fashion strategies to customer involvement, multiple linear regression analysis (with backward approach) is conducted and the results are depicted in Table 9.4.

Observed from Table 9.4, both pricing (beta $= 0.427$, p-value < 0.01) and quick response services (beta $= 0.525$, p-value < 0.01) are found to have significantly positive

Table 9.4 Results of the Multiple Regression Analysis.

Model		Unstandardized Coefficients		Standardized Coefficients		
		B	Std. Error	Beta	t	Sig.
I	(Constant)	.353	.135		2.623	.009
	price	.335	.062	.396	5.449	.000
	design	.074	.067	.072	1.102	.272
	QR	.506	.078	.488	6.527	.000
2	(Constant)	.399	.128		3.124	.002
	price	.362	.057	.427	6.377	.000
	QR	.544	.069	.525	7.832	.000

Dependent Variable: CI
*QR = Quick response
*CI = Customer Involvement

effect on customer involvement (R-square $= 0.873$, p-value < 0.05). On the contrary, the strategy on design is found to have no significant impact on customer involvement with fast fashion brands (p-value of estimated beta > 0.05). The resultant linear regression equation can be established as follows:

$$Y = 0.399 + 0.362X_1 + 0.544X_2, \tag{9.1}$$

where $Y =$ customer involvement, $X_1 =$ pricing, $X_2 =$ quick response service.

Although the three factors are proven to be positively related to customer involvement, the factor of design was excluded from the final regression equation, which reflects that it is relatively insignificant when compared with the other two factors. On the other hand, quick response service is the leading and most significant factor amongst the three in influencing customer involvement as it has the largest standardized coefficient.

From the above finding, we can see that customers emphasize more on the speed of new product launching. Consumers who are fashion conscious are likely to get excited and be happy from finding out new things through shopping (Zuckerman, 1979). In processing information, these customers prefer visual to verbal strategies and are less likely to organize, elaborate, and further analyze the received information (Venkatraman and Price, 1990). Fashion seekers are also perceived to be impulsive and seldom evaluate product carefully (Mittelstaedt et al., 1976). In other words, they would be more concerned about the speed and the variety than the design details of fashion apparels. Thus, our finding that quick response service is the leading strategy to increase customer involvement is consistent with the literature.

9.4.3 Customer involvement and customer attitudes

Correlation analysis is applied to investigate the relationship between customer involvement and customers' attitudes towards fast fashion brands (Hypothesis 2). The results are presented in Table 9.5 below.

Table 9.5 Correlation Analysis – Hypothesis 2.

Correlations

		CI	AL
CI	Pearson Correlation	I	.917**
	Sig. (2-tailed)		.000
	N	200	200
AL	Pearson Correlation	.917**	I
	Sig. (2-tailed)	.000	
	N	200	200

**Correlation is significant at the 0.01 level (2-tailed).

From Table 9.5, the Pearson correlation coefficient between customer involvement (CI) and customer attitudes (AL) is 0.917 (p-value <0.01). This means that customer involvement and customer attitudes are significantly positively correlated and the relationship is strong. Therefore, Hypothesis 2 is supported. In other words, the higher the level of involvement with fast fashion brands, the more positive the attitudes the consumers hold towards fast fashion brands. Observe that a positive attitude towards a brand helps develop important resonance with the brand, and in turn encourages repeat-purchase, which is a pertinent sign of behavioral loyalty. Customers are said to be loyal to a brand if they are willing to spend more time, energy, money and other resources to the brand (beyond those expended during their purchase (Keller, 2001). Brand engagement is hence developed. Therefore, satisfying customers with their needs not only can help fast fashion brands improve their business performance, but also it can promote good word-of mouth recommendations, and strengthen brand image.

9.5 CONCLUDING REMARKS AND IMPLICATIONS

In this paper, we have conducted an empirical study on fast fashion service operations from the consumer perspective. We have shown that the hypotheses are supported. In the following, we discuss the research implications from these hypotheses.

9.5.1 Research implications from Hypothesis 1[2]

From the statistical testing results associated with Hypothesis 1, we have found that there are strong relationships between the strategies of fast fashion and customer involvement. Pricing, product design, and quick response service of fast fashion brands are proven to be attractive to consumers. By comparing amongst the three factors, quick response service is revealed to be the most important characteristic of fast fashion. Some implications from the research findings on Hypothesis 1 are discussed as follows:

---

[2]Including Hypotheses 1a, 1b, and 1c.

(i) Quick response — the agile supply chain system

Quick response is found to be a very important factor for fast fashion. In fact, quick response helps fashion retailers know more about the market. Take Zara as an example, it has established teams of experts who help to seek out new ideas and trends among its target market. The sales floor staff members also help to identify the core preferences of customers. All these vital data and inputs would be sent back to the design team and be quickly converted into tangible products by computer aided design and computer aided manufacturing (CAD/CAM) tools. Thus, under such a quick response system, products would be in the marketplace within a few weeks. Besides having effective ways of gathering market information, to achieve quick response, fashion retailers have to connect and be integrated with their suppliers through sharing point-of-sale data to enhance inventory management. For companies not having vertical integration, improvement of communication between fashion retailers and manufacturers can also be attained through linkage of the ERP systems and other computer based systems between channel agents and establishment of efficient feedback system.

Fast fashion retailers are also suggested to have a network of suppliers which can be flexible enough to deal with sudden changes in demand. As a remark, Zara and Benetton are two fast fashion retailers which work very closely with the network based specialists and manufacturers. In many fast fashion companies, operations (such as dyeing, cutting, labeling and packaging) which can improve efficiency are conducted in-house. All other manufacturing activities, such as the labor-intensive finishing stages, are completed by networks of subcontractors, each of which being specialized in one particular part of the production processes or garment type. This arrangement helps the fast fashion retailers to receive the necessary level of technological, financial and logistical support for better time control and product quality management.

(ii) Quick response implementation — staffing

Teamwork and group problem solving are an integral part of the quick response manufacturing system. Organizing workshops, and giving more autonomy and decision making rights to staff members can enhance these two aspects. These programs can also help boost the staff members' confidence and morale, which are all critical to successful implementation of the quick response supply chain strategy. In addition, improvement programs and reward systems are suggested to be held continuously for helping staff development.

(iii) Pricing strategy — suitable initial price

A suitable initial price is important as it helps attract customers and prevents heavy discounts. In particular, fast fashion products have a very short product life-cycle, and needed to be sold fast. The fast fashion retailers should have a good research on the proper market pricing strategy, with focus on the impact of external factors such as market situation and pricing of their competitors. Setting a good price, supplemented with suitable promotion, is definitely a critical measure. In fact, when customers find that the fashion products create value (with respect to the money they need to pay), they would be more involved.

(iv) Design strategy — intensive research on fashion trends

In order to understand what consumers like, fast fashion retailers should keep up with the fashion trend, make full reference to the related fashion magazines and be aware of the latest fashion shows. All these would give the fast fashion retailers the clue to choose what they should order from the suppliers. Sufficient communication with the suppliers is also essential as they provide an alternative source of information about fashion design trends and strategies.

9.5.2 Research implications from Hypothesis 2

In addition to Hypothesis 1, our findings also support Hypothesis 2 in which there is a strong relationship between customer involvement and customer attitudes towards fast fashion. In other words, if customers' involvement with fast fashion brands increases, they would have a better attitude towards those brands. We discuss the research implications as follows:

Undoubtedly, when consumers have a certain level of involvement with a fashion brand, they are more willing to search for the products they like in the brand; the time and frequency of shopping in that brand will be increased, too. Once customers find the brand is suitable for them, they would prefer it to others and may even be loyal customers to the brand. Fast fashion retailers are hence recommended to focus on the consumers with high customer involvement and offer them some specific promotion schemes and activities. These can further encourage them to be more engaged and stay loyal to the brand. Ultimately, both the brand performance and image would be enhanced.

9.5.3 Research limitations and future research

Similar to other empirical research, there are some limitations in this research study. As time and resources are limited, only the convenience sampling approach with a sample size of 200 is adopted. This may not reflect the real situation of customer attitudes in the market. Besides, the sample comprises more female than male. Thus, a certain level of bias may be introduced. For future research, one can expand the study with a bigger sample size and sampling method with better control of sampling bias. One can also drill deeper and reveal how other important service factors, such as in-store assistance, store atmosphere, etc. in fast fashion retailing may relate to customer involvement and consumer behaviors.

REFERENCES

Attwood, K. 2007. Cheap fashion: the trend may be over. *The Independent*, 3 August 2007.

Barnes, L., Lea-Greenwood, G. 2006. Fast fashioning the supply chain: shaping the research agenda. *Journal of Fashion Marketing and Management: An International Journal*, 10(3), 259–271.

Bruce, G., Daly, L. 2006. Buyer behavior for fast fashion. *Journal of Fashion Marketing and Management* 10(3), 329–344.

Choi, T.M. (Ed.) 2013. Fast Fashion Systems: Theories and Applications, CRC Press.

Christopher, M., R. Lowson, H. Peck. 2004. Creating agile supply chains in the fashion industry. *International Journal of Retail and Distribution Management*, 32(8), 367–376.

Cotte, J. and Wood, S. 2004. Families and innovative consumer behavior: a triadic study of siblings and parents. *Journal of Consumer Research*, 31(1), 78–86.

Hoffman, W. 2007. Logistics get trendy. *Traffic World*, 271(3), 15.

Jackson, T. 2001. The process of fashion trend development leading to a season. *In Fashion marketing: Contemporary issues*, ed. T. Hines, and M. Bruce, 121–132.

Josiam, B. M., Kinley, T. R., & Kim, Y. K. 2005. Involvement and the tourist shopper: Using the involvement construct to segment the American tourist shopper at the mall. *Journal of Vacation Marketing*, 11(2), 135–154.

Jobber, D. 2007. Principles and Practices of Marketing, 5th ed., McGraw-Hill, Maidenhead.

Keller, K. L. (2001). Building customer-based brand equity: A blueprint for creating strong brands.

McKinney, L.N., Legette-Traylor, D., Kincade, D.H., Holloman, L.O. 2004. Selected social factors and the clothing buying behavior patterns of black college consumers. *International Review of Retail, Distribution and Consumer Research*, 14(4), 389–406.

Mittelstaedt, R.A., Grossbart, S.L., Curtis, W.W. and Devere, S.P. 1976. Optimal stimulation level and the adoption decision process. *Journal of Consumer Research*, 3(1), 84–94.

Muncy, J. A., & Hunt, S. D. 1984. Consumer involvement: definitional issues and research directions. *Advances in consumer research*, 11(1), 193–196.

O. Cass, A. 2000. An assessment of consumers, product, purchase decision, advertising, and consumption involvement in fashion clothing. *Journal of Economic Psychology* Issue No. 21, 545–576.

Roehrich, G. 2004. Consumer innovativeness: concepts and measurements. *Journal of Business Research*, 57(6), 671–677.

Slama, M.E., Tashchian, A. 1985, Selected socioeconomic and demographic characteristics associated with purchasing involvement. *Journal of Marketing*, 49(4), 72–82.

Sproles, G., L. Burns. 1994. Changing appearances: Understanding dress in contemporary society. *New York: Fairchild Publications*.

Swinney, R., Cachon, G.P. 2011. The Value of Fast Fashion: Rapid Production, Enhanced Design, and Strategic Consumer Behavior. *Management Science*, 57(4), 778–795.

Venkatraman, M.P. and Price, L.L. 1990. Differentiating between cognitive and sensory innovativeness. *Journal of Business Research*, 20(4), 293–315.

Zaichkowsky, J. 1985. Measuring the involvement construct. *Journal of Consumer Research*, 12(3), 41–52.

Zuckerman, M. 1979. Sensation seeking. *Corsini Encyclopedia of Psychology*.

Subject index

Communications in Cybernetics, Systems Science and Engineering

Book Series Editor: Jeffrey 'Yi-Lin' Forrest

ISSN: 2164-9693

Publisher: CRC Press/Balkema, Taylor & Francis Group

1. A Systemic Perspective on Cognition and Mathematics
   Jeffrey Yi-Lin Forrest
   ISBN: 978-1-138-00016-2 (Hb)

2. Control of Fluid-Containing Rotating Rigid Bodies
   Anatoly A. Gurchenkov, Mikhail V. Nosov & Vladimir I. Tsurkov
   ISBN: 978-1-138-00021-6 (Hb)

3. Research Methodology: From Philosophy of Science to Research Design
   Alexander M. Novikov & Dmitry A. Novikov
   ISBN: 978-1-138-00030-8 (Hb)

4. Fast Fashion Systems: Theories and Applications
   Tsan-Ming Choi (ed.)
   ISBN: 978-1-138-00029-2 (Hb)

5. Reflexion and Control: Mathematical Models
   Dmitry A. Novikov & Alexander G. Chkhartishvili
   ISBN: 978-1-138-02473-1 (Hb)

6. A Systems Perspective on Financial Systems
   Jeffrey Yi-Lin Forrest
   ISBN: 978-1-138-02628-5 (Hb)

7. Fashion Retail Supply Chain Management: A Systems Optimization Approach
   Tsan-Ming Choi
   ISBN: 978-1-138-00028-5 (Hb)

8. Service Supply Chain Systems: A Systems Engineering Approach
   Tsan-Ming Choi (ed.)
   ISBN: 978-1-138-02829-6 (Hb)

Printed and bound by CPI Group (UK) Ltd, Croydon, CR0 4YY

22/10/2024

01777638-0005